IN THE CENTER OF IMMENSITIES

WORLD PERSPECTIVES

Volumes already published

WORLD PERSPECTIVES • *Volume Fifty-three*

Planned and Edited by **RUTH NANDA ANSHEN**

IN THE CENTER
OF IMMENSITIES

Bernard Lovell

1817

HARPER & ROW, PUBLISHERS

New York, Hagerstown, San Francisco, London

IN THE CENTER OF IMMENSITIES. Copyright © 1978 by Sir Bernard Lovell. Introduction copyright © 1976 by Ruth Nanda Anshen. All rights reserved. Printed in the United States of America. No part of this book may be used or reproduced in any manner whatsoever without written permission except in the case of brief quotations embodied in critical articles and reviews. For information address Harper & Row, Publishers, Inc., 10 East 53rd Street, New York, N.Y. 10022. Published simultaneously in Canada by Fitzhenry & Whiteside Limited, Toronto.

FIRST EDITION

Library of Congress Cataloging in Publication Data

Lovell, Alfred Charles Bernard, Sir,
  In the center of immensities.
  (World perspectives; v. 53)
  1. Cosmology. I. Title.
QB981.L87 1978 523.1 76–26241
ISBN 0–06–012716–3

78 79 80 81 82 10 9 8 7 6 5 4 3 2 1

Contents

World Perspectives
What This Series Means

It is the thesis of *World Perspectives* that man is in the process of developing a new consciousness which, in spite of his apparent spiritual and moral captivity, can eventually lift the human race above and beyond the fear, ignorance, and isolation which beset it today. It is to this nascent consciousness, to this concept of man born out of a universe perceived through a fresh vision of reality, that *World Perspectives* is dedicated.

My Introduction to this Series is not of course to be construed as a prefatory essay for each individual book. These few pages simply attempt to set forth the general aim and purpose of the Series as a whole. They try to point to the principle of permanence within change and to define the essential nature of man, as presented by those scholars who have been invited to participate in this intellectual and spiritual movement.

Man has entered a new era of evolutionary history, one in which rapid change is a dominant consequence. He is contending with a fundamental change, since he has intervened in the evolutionary process. He must now better appreciate this fact and then develop the wisdom to direct the process toward his fulfillment rather than toward his destruction. As he learns to apply his understanding of the physical world for practical purposes, he is, in reality, extending his innate capacity and augmenting his ability and his need to communicate as well as his ability to think and to create. And as a result, he is substituting a goal-directed evolutionary process in his struggle against environmental hardship for the slow, but effective, biological evolution which produced modern man through mutation and natural selection. By intelligent intervention in the evolutionary process man has greatly accelerated and greatly expanded the range of his possibilities. But he has not changed the basic fact that it remains a trial and error process, with the danger of taking paths that lead to sterility of mind and heart, moral apathy and intellectual inertia; and even producing social dinosaurs unfit to live in an evolving world.

Only those spiritual and intellectual leaders of our epoch who have a paternity in this extension of man's horizons are invited to participate in the Series: those who are aware of the truth that beyond the divisiveness among men there exists a primordial unitive power since we are all bound together by a common humanity more fundamental than any unity of dogma; those who recognize that the centrifugal force which has scattered and atomized mankind must be replaced by an integrating structure and process capable of bestowing meaning and purpose on existence; those who realize that science itself, when not inhibited by the limitations of its own methodology, when chastened and humbled, commits man to an indeterminate range of yet undreamed consequences that may flow from it.

Virtually all of our disciplines have relied on conceptions which are now incompatible with the Cartesian axiom, and with the static world view we once derived from it. For underlying the new ideas, including those of modern physics, is a unifying order, but it is not causality; it is purpose, and not the purpose of the universe and of man, but the purpose *in* the universe and *in* man. In other words, we seem to inhabit a world of dynamic process and structure. Therefore we need a calculus of potentiality rather than one of probability, a dialectic of polarity, one in which unity and diversity are redefined as simultaneous and necessary poles of the same essence.

Our situation is new. No civilization has previously had to face the challenge of scientific specialization, and our response must be new. Thus this Series is committed to ensure that the spiritual and moral needs of a man as a human being and the scientific and intellectual resources at his command for *life* may be brought into a productive, meaningful and creative harmony.

In a certain sense we may say that man now has regained his former geocentric position in the universe. For a picture of the Earth has been made available from distant space, from the lunar desert, and the sheer isolation of the Earth has become plain. This is as new and as powerful an idea in history as any that has ever been born in man's consciousness. We are all becoming seriously concerned with our natural environment. And this concern is not only the result of the warnings given by biologists, ecologists and conservationists. Rather it is the result of a deepening awareness that something new has happened, that the planet Earth is a unique and precious place. Indeed, it may not be a mere coincidence that this awareness should have been born at the exact moment when man took his first step into outer space.

This Series endeavors to point to a reality of which scientific theory has revealed only one aspect. It is the commitment to this reality that lends universal intent to a scientist's most original and solitary thought. By acknowledging this frankly we shall restore science to the great family of human aspirations by which men hope to fulfill themselves in the world community as thinking and sentient beings. For our problem is to discover a principle of differentiation and yet relationship lucid enough to justify and to purify scientific, philosophic and all other knowledge, both discursive and intuitive, by accepting their interdependence. This is the crisis in consciousness made articulate through the crisis in science. This is the new awakening.

Each volume presents the thought and belief of its author and points to the way in which religion, philosophy, art, science, economics, politics and history may constitute that form of human activity which takes the fullest and most precise account of variousness, possibility, complexity and difficulty.

Thus *World Perspectives* endeavors to define that ecumenical power of the mind and heart which enables man through his mysterious greatness to re-create his life.

This Series is committed to a re-examination of all those sides of human endeavor which the specialist was taught to believe he could safely leave aside. It attempts to show the structural kinship between subject and object; the indwelling of the one in the other. It interprets present and past events impinging on human life in our growing World Age and world consciousness and envisages what man may yet attain when summoned by an unbending inner necessity to the quest of what is most exalted in him. Its purpose is to offer new vistas in terms of world and human development while refusing to betray the intimate correlation between universality and individuality, dynamics and form, freedom and destiny. Each author deals with the increasing realization that spirit and nature are not separate and apart; that intuition and reason must regain their convergence as the means of perceiving and fusing inner being with outer reality.

World Perspectives endeavors to show that the conception of wholeness, unity, organism is a higher and more concrete conception than that of matter and energy. Thus an enlarged meaning of life, of biology, not as it is revealed in the test tube of the laboratory but as it is experienced within the organism of life itself, is attempted in this Series. For the principle of life consists in the tension which connects spirit with the realm of matter, symbiotically joined. The element of life is dominant in the very texture of nature, thus rendering life, biology, a transempirical science. The laws of life have their origin beyond their mere physical manifestations and compel us to consider their spiritual source. In fact, the widening of the conceptual framework has not only served to restore order within the respective branches of knowledge, but has also disclosed analogies in man's position regarding the analysis and synthesis of experience in apparently separated domains of knowledge, suggesting the possibility of an ever more embracing objective description of the meaning of life.

Knowledge, it is shown in these books, no longer consists in a manipulation of man and nature as opposite forces, nor in the reduction of data to mere statistical order, but is a means of liberating mankind from the destructive power of fear, pointing the way toward the goal of the rehabilitation of the human will and the rebirth of faith and confidence in the human person. The works published also endeavor to reveal that the cry for patterns, systems and authorities is growing less insistent as the desire grows stronger in both East and West for the recovery of a dignity, integrity and self-realization which are the inalienable rights of man who may now guide change by means of conscious purpose in the light of the experience of reason.

The volumes in this Series endeavor to demonstrate that only in a society in which awareness of the problems of science exists can its discoveries start great waves of change in human culture, and in such a manner that these discoveries may deepen and not erode the sense of universal human community. The differences in the disciplines, their epistemological exclusiveness, the variety of historical experiences, the differences of traditions, of cultures, of languages, of the arts, should be protected and preserved. But the interrelationship and unity of the whole should at the same time be accepted. For the time-honored dichotomy between science and value judgments must continue to be challenged and we must now require that science tell us not only what *is* but what *ought* to be; to *de*scribe but also to *pre*scribe.

The authors of *World Perspectives* are of course aware that the ultimate answers to the hopes and fears which pervade modern society rest on the moral fiber of man, and on the wisdom and responsibility of those who promote the course of its development. But moral decisions cannot dispense with an insight into the interplay of the objective elements which offer and limit the choices made. Therefore an understanding of what the issues are, though not a sufficient condition, is a necessary prerequisite for directing action toward constructive solutions.

Other vital questions explored relate to problems of international understanding as well as to problems dealing with prejudice and the resultant tensions and antagonisms. The growing perception and responsibility of our World Age point to the new reality that the individual person and the collective person supplement and integrate each other; that the thrall of totalitarianism of both left and right has been shaken in the universal desire to recapture the authority of truth and human totality. Mankind can finally place its trust not in a proletarian authoritarianism, not in a secularized humanism, both of which have betrayed the spiritual property right of history, but in a sacramental brotherhood and in the unity of knowledge. This new consciousness has created a widening of human horizons beyond every parochialism, and a revolution in human thought comparable to the basic assumption, among the ancient Greeks, of the sovereignty of reason; corresponding to the great effulgence of the moral conscience articulated by the Hebrew prophets; analogous to the fundamental assertions of Christianity; or to the beginning of the new scientific era, the era of the science of dynamics, the experimental foundations of which were laid by Galileo in the Renaissance.

An important effort of this Series is to re-examine the contradictory meanings and applications which are given today to such terms as democracy, freedom, justice, love, peace, brotherhood and God. The purpose of such inquiries is to clear the way for the foundation of a genuine *world* history not in terms of nation or race or culture but in terms of man in relation to God, to himself, his fellow man and the universe, that reach beyond immediate self-interest. For the meaning of the World Age consists in respecting man's hopes and dreams which lead to a deeper understanding of the basic values of all peoples.

World Perspectives is planned to gain insight into the meaning of man, who not only is determined by history but who also determines history. History is to be understood as concerned not only with the life of man on this planet but as including also such cosmic influences as interpenetrate our human world. This generation is discovering that history does not conform to the social optimism of modern civilization and that the organization of human communities and the establishment of freedom and peace are not only intellectual achievements but spiritual and moral achievements as well, demanding a cherishing of the wholeness of human personality, the "unmediated wholeness of feeling and thought," and constituting a never-ending challenge to man, emerging from the abyss of meaninglessness and suffering, to be renewed and replenished in the totality of his life.

Justice itself, which has been "in a state of pilgrimage and crucifixion" and now is being slowly liberated from the grip of social and political demonologies in the East as well as in the West, begins to question its own premises. The modern revolutionary movements which have challenged the sacred institutions of society by protecting injustice in the name of social justice are here examined and reevaluated.

In the light of this, we have no choice but to admit that the *un*freedom against which freedom is measured must be retained with it, namely, that the aspect of truth out of which the night view appears to emerge, the darkness of our time, is as little abandonable as is man's subjective advance. Thus the two sources of man's consciousness are inseparable, not as dead but as living and complementary, an aspect of that "principle of complementarity" through which Niels Bohr has sought to unite the quantum and the wave, both of which constitute the very fabric of life's radiant energy.

There is in mankind today a counterforce to the sterility and danger of a quantitative, anonymous mass culture; a new, if sometimes imperceptible, spiritual sense of convergence toward human and world unity on the basis of the sacredness of each human person and respect for the plurality of cultures. There is a growing awareness that equality may not be evaluated in mere numerical terms but is proportionate and analogical in its reality. For when equality is equated with interchangeability, individuality is negated and the human person transmuted into a faceless mask.

We stand at the brink of an age of a world in which human life presses forward to actualize new forms. The false separation of man and nature, of time and space, of freedom and security, is acknowledged, and we are faced with a new vision of man in his organic unity and of history offering a richness and diversity of equality and majesty of scope hitherto unprecedented. In relating the accumulated wisdom of man's spirit to the new reality of the World Age, in articulating its thought and belief, *World Perspectives* seeks to encourage a renaissance of hope in society and of pride in man's decision as to what his destiny will be.

Man has certainly contrived to change the environment, but subject to the new processes involved in this change, the same process of selection continues to operate. The environment has changed partly in a physical and geographical sense, but more particularly from the knowledge we now possess. The Biblical story of Adam and Eve contains a deep lesson, which a casual reading hardly reveals. Once the "fruit of the Tree of Knowledge" has been eaten, the world is changed. The new world is dictated by the knowledge itself, not of course by an edict of God. The Biblical story has further interest in that the new world is said to be much worse than the former idyllic state of ignorance. Today we are beginning to wonder whether this might not also be true. Yet we are uneasy, apprehensive, and our fears lead to the collapse of civilizations. Thus we turn to the truth that knowledge and life are indivisible, even as life and death are inseparable. We *are* what we know and think and feel; we are linked with history, with the world, with the universe, and faith in *Life* creates its own verification.

World Perspectives is committed to the recognition that all great changes are preceded by a vigorous intellectual reevaluation and reorganization. Our authors are aware that the sin of *hubris* may be avoided by showing that the creative process itself is not a free activity if by free we mean arbitrary, or unrelated to cosmic law. For the creative process in the human mind, the developmental process in organic nature and the basic laws of the inorganic realm may be but varied expressions of a universal formative process. Thus *World Perspectives* hopes to show that although the present apocalyptic period is one of exceptional tensions, there is also at work an exceptional movement toward a compensating unity which refuses to violate the ultimate moral power at work in the universe, that very power upon which all human effort must at last depend. In this way we may come to understand that there exists an inherent interdependence of spiritual and mental growth which, though conditioned by circumstances, is never determined by circumstances. In this way the great plethora of human knowledge may be correlated with an insight into the nature of human nature by being attuned to the wide and deep range of human thought and human experience.

Incoherence is the result of the present distintegrative processes in education. Thus the need for *World Perspectives* expresses itself in the recognition that natural and man-made ecological systems require as much study as isolated particles and elementary reactions. For there is a basic correlation of elements in nature as in man which cannot be separated, which compose each other and alter each other mutually. Thus we hope to widen appropriately our conceptual framework of reference. For our epistemological problem consists in our finding the proper balance between our lack of an all-embracing principle relevant to our way of evaluating life and in our power to express ourselves in a logically consistent manner.

Our Judeo-Christian and Greco-Roman heritage, our Hellenic tradition, has compelled us to think in exclusive categories. But our *experience* challenges us to recognize a totality richer and far more complex than the average observer could have suspected—a totality which compels him to think in ways which the logic of dichotomies denies. We are summoned to revise fundamentally our ordinary ways of conceiving experience, and thus, by expanding our vision and by accepting those forms of thought which also include nonexclusive categories, the mind is then able to grasp what it was incapable of grasping or accepting before.

Nature operates out of necessity; there is no alternative in nature, no will, no freedom, no choice as there is for man. Man must have convictions and values to live for, and this also is recognized and accepted by those scientists who are at the same time philosophers. For they then realize that duty and devotion to our task, be it a task of acting or of understanding, will become weaker and rarer unless guidance is sought in a metaphysics that transcends our historical and scientific views or in a religion that transcends and yet pervades the work we are carrying on in the light of day.

For the nature of knowledge, whether scientific or ontological, consists in reconciling *meaning* and *being*. And *being* signifies nothing other than the actualization of potentiality, self-realization which keeps in tune with the transformation. This leads to experience in terms of the individual; and to organization and patterning in terms of the universe. Thus organism and world actualize themselves simultaneously.

And so we may conclude that organism is *being* enduring in time, in fact in eternal time, since it does not have its beginning with procreation, nor with birth, nor does it end with death. Energy and matter in whatever form they may manifest themselves are transtemporal and transspatial and are therefore metaphysical. Man as man is summoned to know what is right and what is wrong, for emptied of such knowledge he is unable to decide what is better or what is worse.

World Perspectives hopes to show that human society is different from animal societies, which, having reached a certain stage, are no longer progressive but are dominated by routine and repetition. Thus man has discovered his own nature, and with this self-knowledge he has left the state of nonage and entered maturity. For he is the only creature who is able to say not only "no" to life but "yes" and to make for himself a life that is human. In this decision lie his burden and his greatness. For the power of life or death lies not only in the tongue but in man's recently acquired ability to destroy or to create life itself, and therefore he is faced with unlimited and unprecedented choices for good and for evil that dominate our time. Our common concern is the very destiny of the human race. For man has now intervened in the process of evolution, a power not given to the pre-Socratics, nor to Aristotle, nor to the Prophets in the East or the West, nor to Copernicus, nor to Luther, Descartes, or Machiavelli. Judgments of value must henceforth direct technological change, for without such values man is divested of his humanity and of his need to collaborate with the very fabric of the universe in order to bestow meaning, purpose, and dignity upon his existence. No time must be lost since the wavelength of change is now shorter than the life-span of man.

In spite of the infinite obligation of men and in spite of their finite power, in spite of the intransigence of nationalisms, and in spite of the homelessness of moral passions rendered ineffectual by the technological outlook, beneath the apparent turmoil and upheaval of the present, and out of the transformations of this dynamic period with the unfolding of a world-consciousness, the purpose of *World Perspectives* is to help quicken the "unshaken heart of well-rounded truth" and interpret the significant elements of the World Age now taking shape out of the core of that undimmed continuity of the creative process which restores man to mankind while deepening and enhancing his communion and his symbiotic relationship with the universe. For we stand on the threshold of a new consciousness and begin to recognize that thought is as powerful an evolutionary force as teeth, claws, and even language.

RUTH NANDA ANSHEN

I

The Dilemma of Mankind

In the thirteenth century Saint Thomas Aquinas wrote his *Summa Contra Gentiles,* endeavoring to establish the truth of the Christian religion. He argued that wisdom was concerned with the end of the Universe; the end of the Universe being the good of the intellect—truth. Therefore the pursuit of wisdom was the most perfect, sublime, profitable and delightful of all pursuits. For another three or four centuries the search for understanding remained largely a philosophical and theological activity on an Earth believed to be conveniently fixed at the center of a Universe whose boundaries were defined by the sphere of the fixed stars. These ideas had emerged more than a thousand years earlier: man's beliefs about the Universe and the physical laws remained substantially the same as those taught by Aristotle in the fourth century B.C.

The strength of Aristotelian physics and astronomy resided in their commonsense attributes, and their ready absorption into the religious doctrines. For an ordinary person it is quite difficult to prove that Aristotle's ideas were wrong for the simple reason that he expressed many of the features of commonplace observations. For us, as for Aristotle, it is the Sun and the stars that rise and set to delineate day from night; unless we are didactic astronomers we continue to use the Aristotelian imagery. As we proceed on our daily tasks it does not immediately appear to us that the Earth is moving at high velocity. If we drop a stone and a feather from a high cliff into the sea, then of course the stone reaches the sea long before the feather. No one, unless possessed of special equipment, could prove the error in the teaching that heavy bodies fall to the ground faster than light bodies, because this is precisely what occurs under ordinary conditions. Indeed, the pleasant story of Galileo and the leaning tower of Pisa is probably wrong since any such demonstration with bodies of significantly dissimilar weights would apparently prove the correctness of Aristotle's teaching: the heavy body would be observed to reach the ground first.

Aristotle's doctrines were a very strong and lasting influence in the history of the world not only because of their compatibility with common observation. Throughout those ages religious dogma was particularly sensitive to any suggestions that the Earth was not fixed at the center of the planets and stars, or that the Universe was not finite in time and space, or that the heavenly bodies did not conform either in motion or shape to the perfection of the circle and the sphere. This ground of compatibility was so strong that no fundamentally contrary views survived. In the second century B.C. Hipparchus discovered the precession of the equinoxes, and Aristarchus of Samos devised an ingenious method for measuring the distance from Earth of the Moon and the Sun. Eclipses of the Sun and Moon could be predicted. There were many such fine astronomical accomplishments, but the concept of the fixed Earth survived.

In the twelfth and thirteenth centuries, with the founding of the universities and the religious orders, a new interest arose in the study of the cosmos. The works of Aristotle were recovered and the interpretation of his doctrines in the hands of Saint Thomas Aquinas continued to provide an acceptable framework for their integration with ecclesiastical doctrine. In the twelve volumes of *Summa Theologiae* Aquinas attacked the forbidding problem of establishing a fabric of understanding which would embrace thirteenth-century cosmology and Christian thought. Throughout the preceding centuries there had been sporadic attacks on the cosmologies bequeathed to the world by Aristotle and Ptolemy, on the grounds of their incompatibility with Holy Scripture. In the emergence of this new age of learning these attacks assumed more menacing proportions. Aquinas was born in 1225, fifteen years after a Provincial Council at Paris had prohibited the teaching of the Greek science and cosmology and ten years after the Fourth Lateran Council had similarly condemned the physics of Aristotle. The triumph of Aquinas was in his recognition that the ancient texts of Aristotle and the Scriptures both required new interpretations and explanations if the mounting conflicts were to be resolved. The thoroughness and skill with which he accomplished this task is remarkable. The cosmological conflicts with the text of Genesis he ascribes to the fact that Moses condescended to the ignorance of the people by speaking to them only of things which were apparent to their senses, and his resolution of the problems of the Ascension is a masterpiece of logic.

The coherence established by Aquinas on the basis of Aristotelian physics and astronomy provided a scheme of great strength for physical and theological teaching throughout the Middle Ages, and attempts to undermine the central features of the doctrine inevitably led to bitter struggles. Indeed, the doctrine may well have continued to survive iconoclastic hypotheses and theories had the intellectual revolution of the sixteenth and seventeenth centuries not changed completely the nature of the pursuit of wisdom as enunciated by Aquinas. For it was the growth of the belief that ideas could be tested and adjusted by experiment and observation that proved fatal to the sublimity of the ancient system of knowledge. For example, a hundred years before the publication of Copernicus's *De revolutionibus,* Nicholas of Cusa had elaborated the idea that the Universe was an infinite sphere, on the argument that no smaller sphere would be consistent with the creative omnipotence of God. It was a mystical idea, concerned with overcoming the paradox that infinite space could be endowed with a central point occupied by the Earth. Cusa's theme had little relevance to cosmology, however, and caused no great disturbance, until interpreted by Giordano Bruno within the framework of the new Copernican doctrine. Bruno's thesis was that a Copernican system of Sun, Earth, and planets need not be at the center of an infinite universe, and, indeed, that the infinity of space could be populated with such systems. Published in three small works in 1584, his ideas were held to be the ideas of a pagan, since they attacked the basic Christian concept of creation. The matyrdom of Bruno—he was burned at the stake in 1600—was the most hideous reaction of a Church whose antagonism to the Copernican doctrine had been unaffected by the ferment of the Reformation. Yet within half a century of Bruno's death the entire scheme of Aristotelian physics and astronomy lay in ruins.

The climax of this great intellectual revolution, in the seventeenth century, is linked with the names of Galileo and Newton. The observations of Galileo provided the evidence in favor of the Copernican heliocentric theory, and Newton's *Principia* supplied the mathematical framework for the concept of the massive Earth in motion around the Sun. The real essence of the revolution, however, resided in Galileo's demonstration that the heavens were neither incorruptible nor unchanging, and in Newton's apparent proof that the Universe was essentially mechanistic. In the thirteen volumes of *Mécanique céleste* Laplace demonstrated that the Universe was a huge machine guided by the Newtonian laws of gravitation and motion. Laplace's demon, knowing the present position and speed of every particle of matter and possessing unlimited powers of calculation, could predict the future for an indefinite time. The need for God in the workings of such a Universe was a belief difficult to sustain.

The totally mechanistic universe of Laplace would have caused Newton to despair, for he had rationalized his position in *Principia* by concluding that "we know Him only by His most wise and excellent contrivances of things and final causes." By the end of the eighteenth century the intellectual revolution had led to this deterministic view of the universe. The long held concept of purpose, which for Aristotle formed an intimate part of scientific procedure and governed the course of development in the Universe, was rejected.

Once more, in the twentieth century, revolutionary changes have been imposed on the foundations of physical theory, undermining the simple mechanistic view. These changes arise primarily from the impact of the quantum theory and the theory of relativity. Both theories have been successful in providing explanations of phenomena which were inexplicable by existing classical theory. The consequences to our understanding of the behavior of matter in the macroscopic world of our daily existence are insignificant. On the other hand in the microworld and for our understanding of the large scale behavior of the Universe the principles of the quantum theory and of general relativity are superior to the classical concepts of Newton and Laplace. Although great successes have been achieved, there remains a fundamental anxiety about the ultimate validity of these theories. The theory capable of the unification of the gravitational and electromagnetic fields has so far evaded us—a failure which, as we shall see, raises acute problems when we attempt to describe the earliest moments in the evolution of the Universe. Indeed, the intractable nature of this problem and the associated difficulties accompanying the attempts to investigate and understand the origins of time and space have led, in many cases, to a revival of elements of transcendentalism in contemporary concepts of the Universe.

The search for a deeper and more comprehensive understanding of the Universe throughout the ages epitomizes the tortuous nature of scientific advance. Astronomical investigations are almost completely observational and not experimental. New ideas, hypotheses and theories about the Universe have tended to survive for long periods because it is not possible to submit them to experimental tests. Perhaps more so than in any other scientific discipline, the majority of new ideas have eventually transpired to be wrong while philosophical concepts with little observational basis have emerged as correct.

It is remarkable that throughout history the individual astronomer, and frequently the community of astronomers, have been convinced that the contemporary view of the Universe has been correct, whereas today we know that a particular view has been either partially or totally incorrect. The complex processes leading to our present understanding of the Universe have led to a modern view of the cosmos which we believe to be substantially correct, but it will be a remarkable and indeed unique feature of human thought if this really is the case. The consequences of this view lead to a intellectual conflict in which the possibility of knowledge and comprehension of the cosmos must be questioned.

The erosion of the fabric of understanding created by St. Thomas Aquinas was a triumph for science, but it created divisions in man's life and thought never subsequently healed. The divisibility of the processes of understanding has led, after only three centuries, to a new dilemma of forbidding proportions. The devices of science, especially those involved with the penetration into the innermost structure of the atom, and the remotest parts of time and space, have led to means of mass destruction never before available to man. This is a special crisis of the modern world, a material dilemma arising from the paradox that the search for knowledge has become dissociated from the search for wisdom. The hope of our generation is to re-establish, as Aquinas did seven hundred years ago, the fabric of man's attempt to comprehend the Universe. It is a compelling task, which leads us to seek the answer to the question raised one and a half centuries ago by Carlyle—What is Man?, ". . . revealed to his like, and dwells with them in Union and Division; and sees and fashions for himself a Universe, with azure Starry Spaces, and long Thousands of Years. . . . as it were, swathed-in, and inextricably over-shrouded; yet it is sky-woven and worthy of a God. Stands he not thereby in the centre of Immensities, in the conflux of Eternities?"[1]

II

The Evolution of Planetary Systems

THE EVOLUTION OF THE HELIOCENTRIC CONCEPT

Today when we are accustomed to the transient nature of
theories it seems remarkable that the fundamentals of Aristotle's physics and astronomy survived for nearly two thousand years, until the age of Copernicus.

It is true that in the fifth or sixth century B.C., Anaximander
believed our world to be only one of many, and that Aristarchus of Samos, who lived from 310 to 230 B.C., advanced the
hypothesis that the Earth and planets moved in circles around
the Sun. But for more than one and a half thousand years after
Aristarchus, our beliefs about the Universe were dominated
by the physics of Aristotle. His treatise *On the Heavens* provided an agreeable description of the cosmological situation,
namely that the spherical Earth was at the center of the Universe and that the heavens were perfectly spherical with the
upper regions more divine than the lower to the extent that
above the Moon everything was indestructible.

Certainly, many difficulties arose. For example, comets had
to be recognized as destructible and had therefore to be assigned to the imperfect sublunary sphere, whereas observational evidence assigned them to places beyond the Moon.
There was, further, a profound theological difficulty. Aristotle
envisaged one unmoved mover giving rise directly to the one

kind of motion which was continuous and infinite—that is, circular motion. The infinite nature of this primary circular motion and the implications of infinite and uncreated time are cardinal to Aristotelian thought. On the other hand the Book of Genesis gives a clear account of the *beginning* of the Universe. In the thirteenth century the conflict between ecclesiastical doctrine and the Aristotelian thesis was brilliantly resolved by Saint Thomas Aquinas. His logical analysis, based on the argument that whatever is moved must be moved by something, and that an endless regression of this type is impossible, was decisive, and led directly to his proof of the existence of God as the unmoved mover.

The most obvious difficulty arising from the Aristotelian insistence on the primary and divine attributes of circular motion was the purely observational one concerning the motion of the planets. Against the background of the "fixed stars" the movement of the planets across the sky appeared to be irregular. The reconciliation of the apparently irregular movement with actual motion in circular orbits became a primary concern of Greek astronomy. About A.D. 150 Ptolemy completed the work of Hipparchus in the construction of a model of the Universe based on circular motions of the Sun and planets around a fixed Earth. The epicycles and deferents of the Ptolemaic universe invoked thirty-nine circular motions in order to explain the observed motion of the Sun, Moon and the five planets then known (Jupiter, Saturn, Mars, Venus, Mercury) as viewed from a fixed Earth. Ptolemy's *Almagest* is a triumph of geometrical constructions based on an entirely false assumption, "that the object which the astronomer must strive to achieve is this: to demonstrate that all the phenomena in the sky are produced by uniform and circular motions." The representation of the motion of the Sun, Moon and planets by uniform circular motion is necessary, Ptolemy explains, "because only such motions are appropriate to their divine nature." Indeed, the epicyclic Universe of Ptolemy was so successful in explaining the observed movements of the planets against the sphere of the fixed stars that it survived for one and a half thousand years.

The revival by Copernicus in the sixteenth century of the concept of Aristarchus, that the Earth was in motion around the Sun, was the beginning of the intellectual revolution which led to the final rejection of the Ptolemaic hypothesis. In modern writings it is often loosely inferred that the new idea of Copernicus marked the final severance with the concept of the perfection of circular motions and with the epicyclic geometry of the Ptolemaic Universe. However, the text of *De revolutionibus orbium caelestium,* published as Copernicus was dying in 1543, lends no substance to this view. On the contrary it seems clear that Copernicus was inspired by the desire to remove a relatively minor defect from the Ptolemaic system. Although the epicycles and deferents of Ptolemy explained the retrogressions in the otherwise eastward movements of the planets across the sky, there remained a difficulty about the uniformity of motion. If the planets were assumed to be in circular motion at a uniform rate about the geometrical center of their circular orbits, then the predictions of the epicyclic theory did not agree exactly with the observed positions. The reconciliation of the predictions and the observations seemed to imply that the Sun and planets moved with different velocities in different parts of their orbits. To overcome this difficulty Ptolemy had introduced the idea of the equant—a point, displaced from the geometrical center of the orbit, about which the motion was uniform. Since the Sun or a planet moves at a uniform rate with respect to the equant point then clearly, when viewed from the geometrical center of the orbit, the planet seems to move at an irregular rate, or to wobble.

In the preface to *De revolutionibus* addressed to Pope Paul III Copernicus writes ". . . those again who have devised eccentric systems [involving equants], though they appear to have well-nigh established the seeming motions by calculations agreeable to their assumptions, have yet made many admissions which seem to violate the first principle of uniformity in motion." By affirming that it was the Earth that was in motion around the Sun Copernicus eliminated the need for equants and thus established a major simplification in the Ptolemaic scheme. However, he did not propose that the motion of the planets was also Sun centered. On the contrary, his supposition was that the planets moved on epicycles with respect to the orbit of the Earth. The Copernican system required at least as many epicyclic motions as the Ptolemaic system; neither did the predictions of his heliocentric theory give better agreement with the observations.

The Copernican hypothesis restored the perfection of the uniformity of circular motion at the cost of raising the immense problem of the motion of the Earth, an idea in conflict with both common sense and theological belief. The proposition met with objections of increasing antagonism for the next seventy or eighty years. Even four years before the publication of *De revolutionibus* Martin Luther had said, "People gave ear to an upstart astrologer who strove to show that the Earth revolves, not the heavens or the firmament, the Sun and the Moon. . . . This fool wishes to reverse the entire science of astronomy; but sacred Scripture tells us that Joshua* commanded the sun to stand still and not the earth."[1]

Six years after the death of Copernicus the influential academic and Protestant reformer, Philip Melancthon, challenged the idea that the Earth was in motion by referring to the witness of the eyes that the heavens revolved in the space of twenty-four hours: ". . . but certain men, either from the love of novelty, or to make a display of ingenuity, have concluded that the Earth moves. . . . it is a want of honesty and decency to assert such notions publicly, and the example is

---

*Luther's reference is to Joshua 10:12,13.
12. Then spake Joshua to the Lord in the day when he had delivered up the Amorites before the children of Israel, and he said in the sight of Israel, Sun, stand thou still upon Gibeon, and thou Moon, in the valley of Ajalon.
13. And the Sun stood still, and the Moon stayed, until the people had avenged themselves upon their enemies. Is not this written in the book of Jasher? So the Sun stood still in the midst of heaven, and hasted not to go down about a whole day.
There is also a biblical reference to the Sun standing still in Habakkuk 3:11.
The sun and the moon stood still in their habitation; at the light of thine arrows they went, and at the shining of thy glittering spear.

pernicious. It is part of a good mind to accept the truth as revealed by God and to aquiesce in it."[2]

In his detailed analysis of such objections, T. S. Kuhn remarks that "by the first decade of the seventeenth century clergymen of many persuasions were to be found searching the Bible line by line for a new passage that would confound the adherents of the Earth's motion."[3] Nevertheless, it was over this period that the transition from the geocentric to the heliocentric view became inevitable, a transition associated with the genius of three men: Tycho Brahe, Johannes Kepler and Galileo.

It is ironical that Tycho Brahe, who lived from 1546 to 1601, was a powerful and influential opponent of the Copernican hypothesis, for he revolutionized observational astronomy. As distinct from the observers of antiquity he made *systematic* observations of the stars and planets, and with an accuracy never before achieved. He soon discovered that there were large errors in the existing ephemerides—for example, that the predicted time of the conjunction of the planets Saturn and Jupiter in 1563 was a month wrong according to the Alfonsine numbers and even several days wrong according to the Copernican tables. Tycho Brahe devised his own system of epicyclic motions to account for the motions of the heavenly bodies, but his system was a regression, an uneasy compromise in which he reinstated the Earth as the center of the universe, with the planets circling around the Sun and the Sun around the Earth.

One of the cardinal events in the history of astronomy is the quarrel between Tycho Brahe and Christian IV, who in 1588 had succeeded Tycho Brahe's sponsor, Frederick II, as King of Denmark. For most of his life, Tycho Brahe lived and worked under sumptuous conditions at Uranieborg on the island of Hven, which lies in the sound between Denmark and Sweden and is now Swedish territory.* In 1597, however, Tycho Brahe departed from Hven and in 1599 reached

---

*The foundations of Tycho Brahe's castle at Uranieborg and the reconstructed buildings of his observatory can still be seen by visitors to this island.

Prague with all his instruments and library. In a castle near Prague he began to reassemble his instruments, and although he made no more observations during the few years before he died, those years were of immense significance, for he met Kepler, who at last came into possession of Tycho Brahe's accurate, systematic measurements of planetary positions.

Kepler, born in 1571, appears never to have doubted the correctness of the Copernican heliocentric hypothesis. Before he met Tycho Brahe he had already made many improvements in the heliocentric system of Copernicus without departing from the concept of circular motion, eliminating a number of problems common to both the Ptolemaic and Copernican systems. For example, Copernicus had retained the Ptolemaic idea that all the orbits of the planets must intersect at the center of the Earth. Kepler maintained that since the Copernican system was Sun-governed, the orbits should intersect in the Sun.

There remained the problem of the planet Mars, whose apparent irregular motion had always presented great difficulty. For many years Kepler had tried various combinations of circles, and his solutions fell within the limits of ancient observations. But compared with the more precise observations of Tycho Brahe they were in error by eight minutes of arc. Eventually Kepler concluded that no possible system in which the planets moved in circular orbits could agree with the observations, and in the historic publication of 1609, *On the Motion of Mars,* he showed that theory and observation could be reconciled if the planets moved in elliptical orbits with the Sun at one of the foci of the ellipse, and if the speed of the planet along the ellipse was such that the line joining the Sun and planet swept out equal areas in equal times. Ten years later in *Harmonies of the World* Kepler enunciated his third law of planetary motion, relating the speed of planets in different orbits: the ratio of the squares of the orbital periods of two planets is equal to the ratio of the cubes of their average distances from the Sun.

In the year when Kepler published his first and second laws

of planetary motion Galileo first looked at the heavens through a small telescope. With this he observed the moons of Jupiter, the phases of Venus, the spots on the Sun and the cratering on the Moon. The observational evidence for the heliocentric concept became overwhelming and the belief in the perfection of celestial bodies was undermined.

Copernicus, Tycho Brahe, Kepler and Galileo were the four men responsible for the overthrow of the geocentric doctrine. Galileo suffered the full impact of the opposition. In fact, on 11 November 1572, forty years before Galileo published the results of his observations, Tycho Brahe had observed a new star as bright as Venus, near the constellation of Cassiopeia, where no bright star had been observed before. He had determined its position and proved that it was motionless among the fixed stars—in a region hitherto regarded as fixed and unchanging in accordance with Aristotelian doctrine.* But apart from this observation, nothing previously had cast doubt on the perfection of the heavenly bodies beyond the sublunary sphere, and the heliocentric doctrine could still be regarded as a mathematical device. In 1610 with the publication of Galileo's *Siderius Nuncius* the opposition reached fanatical proportions. The Catholic Church officially joined the battle to preserve the ancient doctrines and the charge against the heliocentric concept became one of formal heresy, and *De revolutionibus* was placed on the index of forbidden books in 1616. The battle, however, was already lost. The trial and judgment of Galileo[4] marks the climax of the century spanning the death of Copernicus in 1543 to that of Galileo himself in 1642, during which a major revolution occurred in science and cosmology and in Western Man's theological and intellectual attitude toward the cosmos.

---

*Now we understand that Tycho Brahe had observed one of the three supernova explosions in the recorded history of the Milky Way. The other two are the Crab nebula, or the supernova event recorded by Chinese astronomers in A.D. 1054, and the event recorded by Kepler in A.D. 1604.

THE PHYSICAL SYNTHESIS

The establishment of the heliocentric concept created a severe intellectual dilemma, the pattern of which we recognize again today in the human conflicts engendered by our contemporary science. It is nearly three centuries since the conflict aroused by the heliocentric theory was substantially resolved and yet even today it is not difficult to feel sympathy with those who resisted the new science.

The heliocentric theory did possess marginal advantages over the Ptolemaic system for the astronomers of that age, but for what appeared to be a minor improvement in the explanation of the movement of the heavenly bodies, a staggering blow was delivered to theological dogma which had survived in harmony with science for nearly two millennia. One has merely to look at the heavens on a clear night to feel the force of Melancthon's argument in 1549 that "the eyes are witnesses that the heavens revolve in the space of twenty-four hours." This was a major difficulty of common sense. After all, the size of the Earth had been known for almost two thousand years: in the third century B.C. Eratosthenes had used a vertical gnomon at Alexandria to determine the inclination of the noon Sun when it was directly overhead at Syene, 5,000 stades south of Alexandria, and from this had estimated the Earth's circumference to be 250,000 stades—a value only a few percent less than the modern value of 24,000 miles. The idea that such a vast globe inhabited by man was in motion naturally seemed to be in conflict with reason. And even though the physical characteristics of the Earth and its motion through space have long ago been absorbed into human thought, nevertheless, as we crawl in a traffic jam into a modern city, it is still hard to envisage that for every minute of our slow progress we have simultaneously been moved through space over 1,000 miles because the Earth is moving around the Sun, and that we have been translated another 17.4 miles because of the rotation of the Earth

—and that this motion is of a globe with a mass of a billion trillion tons. Those who objected to the heliocentric idea were quite right to ask what could maintain this annual and diurnal motion of the Earth on which they lived in an apparently stationary condition.

In the Aristotelian and Ptolemaic world view the answers about the motion of the heavenly bodies were concise. The stellar sphere carried the stars in their diurnal motion and this sphere was the driving force for the motion of the planets. The Universe was finite, defined by the outer stellar sphere, and the center of the sphere was the absolute center of space. The absolute center was occupied by the Earth, and all bodies aspired to reach the center of the Earth because it was the center of the Universe. The heliocentric Universe of Copernicus and Galileo destroyed the basis of these beliefs but the physics of their epoch could offer no alternative explanation for the celestial motions nor for the obvious fact that stones fell toward the Earth, although it was in motion and not the center of the Universe. These intellectual distractions following the transition from the geocentric to the heliocentric world view illuminate the genius of Isaac Newton, born in the year of Galileo's death.

Of course the physical problem of planetary motions in the heliocentric system had throughout been a major concern. In his work published in 1600, *On the Magnet,* William Gilbert had recognized that the Earth was a huge magnet; Kepler, generalizing this to include the Sun and planets, envisaged that the planets were moved in some way by forces emanating from the Sun—a mixture of *anima motrix* (a system of rays emanating from the Sun), which produced the circular motion, modified into elliptical motion by a second force identified as magnetism. The views of the French philosopher René Descartes were of great significance in the development of these ideas. In *Principles of Philosophy,* published in 1644, Descartes enlarged upon the concept that the motions of particles were governed by laws imposed by God at the Creation. In a contemporary version of this medieval idea he gave the first clear statement of inertial motion, namely that a particle at rest in a void will remain at rest, and that one in motion will remain in motion in a straight line at the same speed unless deflected by another particle. He maintained that since, in nature, particles in general changed their speed and direction, so these changes must arise from forces exerted by other bodies.

This recognition of the linearity of unimpeded motion was immensely important. The concept was subsequently applied to the problem of planetary motion by both Giovanni Borelli and Robert Hooke. Borelli recognized the need for a central force in order to maintain a body in orbit around the Sun, while in 1666 Hooke clearly demonstrated the equivalence of terrestrial and celestial motion during a lecture to the Royal Society, using a pendulum bob suspended so that it could move freely in any direction. By imparting various initial speeds and directions to this bob he showed that it would move in a circular or elliptical orbit about a central point and argued that similar forces in the heavens determined the orbital motion of the Earth and planets around the Sun.

The conclusion that the same force of attraction governed the motion of bodies on Earth and of the planets around the Sun was of immense importance in the development of science and cosmology. Whether Hooke or Newton first reached this revolutionary conclusion is unknown. The concept in its qualitative form was clearly enunciated by Hooke in a publication of 1674, although it seems likely that Newton had independently arrived at a similar conclusion.

The controversy with Hooke cannot obscure the fact that it was Newton who gave mathematical substance to these ideas. He deduced the rate of apparent fall of a planet to the Sun, or of the Moon to the Earth, if they were to remain in stable orbits. Assuming that Kepler's third law governed the geometry of the planetary system, Newton deduced that the force of attraction between the planets and the Sun must vary inversely as the square of the distance separating them from the Sun. He then recognized that the same inverse square law governed the motion of the Moon and of a stone falling to Earth. With the publication of *Principia Mathematica* in 1687 the physical synthesis was complete. The heliocentric concept and the Keplerian laws of planetary motion appeared as a natural consequence of a fundamental law of nature—the inverse square law of gravitation.

The publication of *Principia* marks the beginning of a comparatively stable period in cosmological thought. The success of the inverse square law of gravitation tended to obscure the fundamental problems associated with the nature of the gravitational force—and of the initial cause of motion in the Universe. The idea of action at a distance through a vacuum presented a great difficulty for Newton himself, but since no physical explanation could be found and because the gravitational theory seemed complete mathematically, the idea of gravity as an intrinsic property of matter was gradually accepted and remained unchallenged until the publication of Einstein's general theory of relativity in 1916. For over two hundred years the system of Sun, Earth and planets was believed to exist at the center of an absolute but infinite space in which the motions of the particles were governed by the laws of inertia and gravity. The heliocentric concept became embedded in an egocentric belief that the planetary system inhabited by man was at the center of the cosmos. On the whole these were centuries of a comforting philosophical stability in which the Universe was envisaged as made for man by God, who had in the beginning endowed the particles of the Universe with such properties as were necessary for the appropriate inertial motions and gravitational forces.

THE ORIGIN OF THE PLANETARY SYSTEM

Since the acceptance of the Newtonian concepts the attention of astronomers has on the whole been directed to other problems. The major difficulties surrounding the motion of the planets had been resolved, the solution being based on the five planets known since ancient times. A sixth planet,

Uranus, was discovered by William Herschel in 1781. In 1845 J. C. Adams of Cambridge and U. J. J. Le Verrier in Paris independently predicted that certain irregularities in the motion of this planet must be caused by the presence of another unknown planet. Their predictions led to the discovery of Neptune in 1846. Percival Lowell, the founder of the observatory at Flagstaff, Arizona, began the search for a trans-Neptunian planet in 1905 on the basis of calculations which he published in 1914. Lowell died in 1916, but the search was resumed in 1929 using a new 13-inch telescope. On 12 March 1930 the observatory announced the discovery of the last of the known planets, Pluto, by a young assistant, Clyde Tombaugh.

Although for this system of planets Newtonian theory provides an almost precise description of planetary motions, some problems remained. The most important of these was that the motion of the perihelion of the planet Mercury could not be explained by Newtonian theory. This problem was resolved in 1919 when Einstein showed that the displacement of the perihelion of the planet by about forty-three seconds of arc per century was in accordance with the predictions of general relativity, and today it seems that the general theory of relativity—in which Newtonian theory appears as a special case mathematically but not, of course, conceptually—probably gives a precise description of planetary motions. It is true that attention has been given in recent years to small differences between the predicted and measured positions of Neptune since its discovery in 1846; some astronomers believe that the orbit of the planet may be influenced by another planet yet to be discovered.

The main concentration of interest today is, nevertheless, on the physical condition of the planets and the problem of their origin and evolution. The disparity of views is remarkable. Although there have been periods during the last century when some degree of coherence of ideas about the origin of the system seemed to be emerging, a quarter of a century ago one of the great authorities on these problems said in a lecture to the Royal Society that "There is hardly a feature of

our system that I would regard as satisfactorily explained."[5]
That same remark could be made today.

Historians of this subject trace the beginning of serious
consideration of the formation of the solar system to Descartes. In 1644 his remarkable intuitive ideas about inertial
motion were extended to the concept that the primeval cor-
puscles would ultimately circulate in vortices filling the Universe and that such vortices were potentially solar systems. A
more fruitful concept was enunciated a century later by Kant.
In his *General Natural History and Theory of the Heavens* (1755),
he suggested that the Universe was almost uniformly filled
with gas and that the solar system formed by condensation
and subsequent contraction around a region of slightly higher
density. A more definitive idea of this type was proposed by
Laplace in 1796, and there are many who believe today that
his nebular theory may well be correct in principle, if not in
detail.

The hypothesis of Laplace and all the variations of it which
have followed make the important assumption that the planets were formed from a diffuse cloud of gas surrounding the
Sun. Laplace believed that the planetary system condensed
from a rarefied nebula of gas in slow rotation. As the nebula
condensed the angular momentum increased so that the
cloud assumed a shape like a lens. The internal gravitational
forces causing the contraction were eventually balanced by
the centrifugal forces of the rotating nebula. At that stage the
rotation was assumed to be so rapid that some of the gas was
ejected and formed a ring outside the nebula. This process
happened repeatedly until the central region of the nebula
condensed into the Sun, and planets formed in each of the
ejected rings.

These ideas of Laplace accounted well for the general features of the solar system and enjoyed a long period of success,
but toward the end of the nineteenth century they were subjected to criticism. For example, Clerk Maxwell proved that
the material in the rings of detached gas would not coalesce
into planets but would be transformed into a series of much
smaller bodies like Saturn's rings. The theory also faced the

serious criticism that it could not account for the unusual distribution of mass and angular momentum in the solar system. The problem was that the Sun contains nearly all of the mass in the solar system, being about 750 times more massive than the whole of the remaining planetary system; therefore, the rotational momentum of the system should be concentrated in the Sun as well. In fact the Sun is rotating slowly compared with the planets and the ratio of the angular momenta of the planets to the Sun is about 200 to 1. On the simple theory of Laplace the distribution of mass and momenta in the system must have occurred at the time of ejection of the peripheral rings from the contracting nebula. If such a minor part of the mass of the nebula separated, possessing the greater part of the angular momentum, then very high velocities must have been involved and condensation of the material of the rings into planets could not have taken place.

This dynamical problem was undoubtedly the stimulus leading to a number of different ideas in the early years of this century. Independently, the American astronomers T. C. Chamberlin (in 1901) and Forest Moulton (in 1905) suggested that the Sun formed originally as an ordinary star without planets and that the planets were formed as a result of a subsequent encounter with another star. They suggested that because of the gravitational attraction between the Sun and this star great amounts of gaseous matter were torn from the Sun, some of which remained under the gravitational attraction of the Sun, condensing into small fragments and finally accreting into larger bodies to form the planets.

The British astronomers James Jeans (in 1916) and Harold Jeffreys (in 1918) proposed variations of this encounter hypothesis. In Jeans's theory the encounter did not directly tear matter from the Sun, but produced a tidal bulge from which the planetary material was ejected. Jeffreys suggested that in a grazing encounter fluid filaments would be formed which would break away from both the Sun and the star. The material in the filaments would then break up into droplets and eventually condense into the planets. The encounter theories received a severe setback in 1935 when H. N. Russell showed that in any such encounter the star would have to pass so close to the Sun that the planets would eventually move in orbits thousands of times closer to the Sun than those which exist.

During the last fifty years there have been many variations on these theories. In the decade from about 1936 to 1946 the ideas of Fred Hoyle and Raymond Lyttleton achieved a certain popularity. According to these theories the Sun was originally a member of a binary or triple star system and the planets were formed after a breakup of the companion star resulting either from rotational instability in the companion star, or (according to Hoyle) because the companion underwent a supernova explosion. When contact with Soviet astronomers was re-established after World War II it was found that Otto Schmidt and B. J. Levin had developed comprehensive theories of a different type. The Soviet school believed that the Sun was first an ordinary star which in its subsequent journey through interplanetary space moved into a cloud of dust and gas, capturing some of it to form a solar nebula from which the planets were formed.

The recognition during the last ten or twenty years that magnetic fields must have a significant rôle in the Universe has provided an escape from the problem of the mass-momentum distribution in the solar system: it is argued that the unusual distribution could result from a magnetic coupling between the Sun and the planetary disk. At the same time, as our knowledge of the physical condition of the planets has increased, other problems have arisen. For example, any theory must explain why the four planets nearest to the Sun (the terrestrial planets—Mercury, Venus, Earth, Mars) are over two hundred times less massive than the outer major planets (Jupiter, Saturn, Uranus, Neptune).* Furthermore, although the outer planets contain this preponderence of the total mass of material, their densities are, on average, two to three times less than those of the terrestrial planets. This is a conse-

*Mercury, Venus, Earth, Mars have a mass 1.97 times the mass of the Earth, while Jupiter, Saturn, Uranus, Neptune have a mass 444.74 times the mass of the Earth.

quence of a radically different composition: whereas the terrestrial planets are predominantly composed of heavy, nonvolatile material, the major planets have a composition similar to that of the interstellar material from which the Sun was formed.

Of the several dozen theories published during the last few decades and not yet finally discarded, contemporary opinion would seem to be in favor of some modern version of the Laplace nebular hypothesis. As for the process by which the Sun achieved the nebula from which the planets formed, opinion today tends to the view that the Sun and planets were formed at the same time from an interstellar cloud. Laplace spoke in terms of the condensation of planets from the gaseous nebula. Today it is commonly believed that processes involving the collision of dust and gas particles in the nebula led to the growth of the planets—a process which might occupy a few thousand million years. The cratering so evident on the Moon, Mars and Mercury might well be the result of the violent collisions to be expected in the later stages of the accretion of such bodies.

The opinion that the solar nebula was part of the gas cloud from which the Sun itself condensed is supported by modern photographs of the gaseous nebulae in the spiral arms of the Milky Way. In the well-known nebula in the constellation of Orion, for example, photographs reveal large numbers of dark globules—and the evidence is that these globules are the early stages of stars forming from the material of the nebula. Mathematical treatment of the details with the help of modern computers suggests that the evolutionary process may occur in the following way. If the interstellar cloud is large enough and of sufficient density then, with the atoms in random motion, condensed pockets of gas may arise. These pockets will contain so many atoms that the condensation is preserved by self-gravitation. To give rise to such a contracting globule an immense number of atoms is required, about 10^{57}. It is a number difficult to visualize—it has been estimated that a count of all the grains of sand on the world's beaches would only reach about 10^{25}. Nevertheless, the number is reasonable because if the atoms are hydrogen then the mass of the 10^{57} atoms would be some 10^{33} grams, which is about the mass of an average star. (The mass of the Sun, for example, is 1.98×10^{33} grams.) The computations show that once a condensation of this type begins, events happen quickly as far as cosmic time scales are concerned. Initially the diameter of the globule may be a few trillion miles but this decreases rapidly to a few hundred million miles. After about 27 million years the contraction will have increased the pressure in the globule to several thousand million atmospheres and the temperature will have risen to about 20 million degrees—conditions under which the thermonuclear transformation of hydrogen to helium occurs. The release of nuclear energy would generate enough outward pressure to prevent further contraction of the globule, and in fact the globule at this stage will have reached the stable condition of a star like the Sun in which during every second some 564 million tons of hydrogen are transformed into 560 million tons of helium. This rate of conversion of mass into energy in the Sun represents conversion of only one tenth of one percent of its mass every ten thousand million years, so that these processes, which have been operating for at least 4 billion years, can continue for several billion years more before the conditions of stability are disturbed. A most significant aspect of this contemporary view of the birth of stars is that they condense in large numbers from a huge interstellar cloud and in this process the remnants of the interstellar cloud will be retained around the stars as nebulae from which planetary systems will form—just as the solar system formed from the nebula around the Sun.

In recent years a very large number of scientific papers has been published on this question of the origin and evolution of the solar system. If one considers the system as an isolated unit in the Universe then it is possible to specify a number of criteria which any theory must satisfy. The most prominent is that the theory must be able to account for the peculiar circumstance that the Sun, which contains 99.9% of the mass of the system carries only 2% of the rotational momentum. The theory must also explain the striking difference in mass, density and constitution of the inner terrestrial planets compared with the outer major planets. As regards the first difficulty a further problem arises if the scope of the enquiry is widened to take account of the contemporary observational evidence regarding the formation of stars in the gas clouds of the Milky Way.

These gas clouds in the spiral arms share in the general rotation of the galaxy and thus some process must lead to the removal of angular momentum from the Sun and the stars. This presents a more serious difficulty for the encounter-type theories than for the theories postulating that the planets originated from a solar nebula which was originally part of the gas cloud—and that the angular momentum is largely taken up by the planets forming in such a nebula. As regards the second difficulty of the radically differing constitution of the inner and outer planets this also seems to be a more formidable problem on the encounter theories than on the nebular hypothesis.

From these various considerations contemporary opinion inclines to accept a nebular theory for the origin of the planetary system, and it is believed that the Sun acquired the nebula as a natural consequence of its condensation from the primeval gas cloud. If this view is correct then a most important consequence is that planetary systems around stars must be a common feature of the Universe—and thus the solar system can no longer be regarded as effectively unique because of the extreme rarity of stellar encounters. A host of difficulties and uncertainties remain, not only in regard to the detailed processes by which planets form from a solar or stellar nebula, but also because there is not yet decisive observational evidence for the existence of planetary systems around stars other than the Sun. Nevertheless the circumstantial evidence, both for the processes of stellar formation from the interstellar clouds and for the existence of planets around some of the nearer stars, is considerable. A Universe in which there exists a multiplicity of planetary systems has become a common feature of contemporary astronomical thought, and there are many who believe also that it is not unreasonable to consider the possibility that organisms may have developed in planetary systems other than the solar system.

III

Investigation of the Solar System

When considering the progress made in the investigation of the solar system it is important to remember that contemporary technology has influenced these inquiries only in recent years. The eyes of the ancient investigators were assisted only by elementary devices like the gnomon or the sundial. As we have seen, in the latter part of the sixteenth century observational astronomy was revolutionized by Tycho Brahe. When he was a young man, in his early twenties, he built near Augsburg a great quadrant of brass and oak, eighteen feet in radius, and although it was exposed to the elements he used it to measure the altitude of the Sun and planets to one sixth of a minute of arc. When Tycho Brahe returned to Denmark after completing his studies in the European universities, the mayor of Augsburg continued to make observations with the quadrant and sent the results to Tycho Brahe. Unfortunately the instrument soon became derelict, and to those who might despair of the contemporary attitude to science it is interesting to read that four hundred years ago Tycho Brahe wrote:

> One could have wished that this excellent instrument had been preserved for a longer period in this place, and had stayed in use, or, else, that another instrument had been constructed in its place. Since, however, men as a rule are more interested in worldly matters than in things celestial, they usually regard with indifference such happenings which will perhaps be more harmful to them than they themselves realize.

The quadrant at Augsburg was soon replaced by the series of great quadrants and circular equatorials which Tycho Brahe built at his observatory on the island of Hven. With these he made observations of the heavenly bodies which were both systematic and far more precise than any made previously. We have seen how these measurements provided the decisive link between the heliocentric hypothesis of Copernicus and the Keplerian laws of planetary motion.

Tycho Brahe occupies a unique place in the history of astronomy. He may not have been the last of the great naked-eye astronomers, but he was certainly the last before technical developments began to give important assistance to the eye. For within a few years of his death a telescope was used by Galileo to study the heavens.

Galileo did not invent the telescope; the Dutch spectacle-maker Johann Lippershey claimed a license in October 1608 for the manufacture of telescopes with single and double lenses.[1] It is not even certain that he was the first to use the new type of instrument for observing the heavens, but he was certainly the first person to do so with an understanding of the significance of his observations. Two of Galileo's telescopes preserved in Florence have lenses with diameters of 4 and 4.4 centimeters and focal lengths of 95 and 125 centimeters. Nevertheless, he claimed in *Siderius Nuncius* that he had succeeded with these small lenses in constructing an instrument so superior "that objects seen through it appear magnified nearly a thousand times, and more than thirty times nearer than if viewed by the natural powers of sight alone."

For nearly three and a half centuries from the time of Galileo to the mid-twentieth century when, through the medium of the new science of radio astronomy, it became evident that important information about the Universe could be obtained in regions of the spectrum lying outside the narrow visual range of wavelengths, the development of the optical telescope was the major feature in astronomy. Throughout this period optical telescopes of ever-increasing size and accuracy were constructed, culminating in the Soviet telescope with a mirror diameter of 236 inches, built in the North Caucasus and commissioned in 1976. Throughout this period, too, important auxiliary instruments were developed for use with the optical telescope. In 1814 Fraunhofer discovered the spectral lines in the light from the Sun; thereafter spectroscopes became an essential part of astronomical telescopes and the new science of astrophysical spectroscopy was initiated. The first use of photography with telescopes can be traced back to about 1840 but it was 1876 before dry plates were used. Subsequent improvements in photographic techniques led to significant increases in the sensitivity of telescopes and today the application of electronic and digital recording techniques is still further enhancing the sensitivity of the optical telescope.

Immense contributions have been made to our knowledge of the Universe by these telescopic systems, but the increase in our knowledge of the solar system has been less dramatic. This somewhat strange situation has occurred because, as far as the solar system is concerned, the achievement of maximum sensitivity in the telescope system has not been the prime requirement. The real difficulty is that the turbulence in the Earth's atmosphere even at the best sites sets a limit of about 0.5 seconds of arc to angular resolution. The 200-inch telescope on Palomar Mountain in California has a theoretical resolution of a few hundredths of a second of arc, but in practice the resolution is limited to that which would be obtained with a telescope of about one tenth the size. This means that severe limitations have been placed on the study of the physical condition of the Moon and planets from Earth. For example, at its mean distance from Earth the Moon subtends an angle of 31 minutes of arc, so that even with the largest telescopes, since the resolution is no better than 0.5 seconds of arc, a feature on the lunar surface must be about a kilometer in extent before it is resolved. For Mars at its closest approach to Earth the corresponding size of a feature on the planet would be about one hundred thirty kilometers and for Venus, ninety—but in that case the planet is perpetually covered in cloud and the surface can never be seen from Earth. Because of these constraints, until the mid-twentieth century only the basic physical elements of the planets were known, that is, their size, mass and orbit, and what were believed to be reliable measurements of their surface temperatures. Spectroscopic studies had also led to assessments of the proportion of various gases in the planetary atmospheres. Even some of these basic facts were unknown for Venus, since the clouds surrounding the planet were featureless, and although many theoretical papers had been published on such problems as the rotation rate of the planet on its axis, there were very wide disagreements.

The situation was considerably improved by the advent of radio and radar techniques. With the use of radio telescopes it became a relatively straightforward matter to measure the intensity of the radiation from a planet at wavelengths in the centimeter band and thereby to refine the values for surface temperatures. The development of radar techniques led to an even more important advance. With a powerful transmitter and large radio telescope, it became possible to observe a radar beam transmitted from Earth after reflection from the planet Venus. The timing of the interval between transmission and reception enabled the distance of the planet from Earth to be measured precisely for the first time.

The lack of real precision in our knowledge of planetary distances until recent times is surprising. For over three hundred years the Keplerian laws were used to calculate the position of the planets in the sky and their distances from Earth. These calculations depend on the knowledge of the solar parallax—that is, the angle subtended by the radius of the Earth at the Sun. Over the centuries various methods had been used to measure this important parameter. From Halley's attempt to do this by observation of the transit of Venus across the solar disk in 1716 until the mid-twentieth century, the most accurate measurements differed among themselves by 0.1 percent. When the Soviet Union launched the first space probe to the planet Venus, on 12 February 1961, the distance of Venus was not known with sufficient accuracy to make contact between space probe and planet certain. However, the transmitter controlling the probe could also be used to measure the planetary distance by radar and in that year three independent radar measurements from the USA, the USSR and the UK achieved success. With these radar techniques the astronomical unit (the mean distance of the Earth from the Sun) was derived to an accuracy of 1 part in 30 million. Similar radar systems soon settled the problem of how fast Venus was rotating on its axis—existing estimates varied from 24 hours to 225 days. During the year 1963–64 the analysis of the spread in frequency of radar waves reflected from Venus gave a decisive answer: the planet was rotating on its axis with a period of 244.3 days and furthermore the rotation was retrograde.

Mercury, like Venus, also presents particular difficulty for optical observers because its angular separation from the Sun is never greater than twenty-seven degrees and it is therefore never seen against a dark-sky background. The belief that Mercury always presented the same face to the Sun, based on the assumption that it rotated on its axis only once during its 88-day orbital period, was disproved by radar measurements in 1965. The axial rotation period is 58.6 days, so that there is no region of permanent day or night on the planet, although the length of the Mercurian day, which is equivalent to 176 of our terrestrial days, means that the sunlit surface gets very hot—about 600°K*—while the long nights are extremely cold—about 100°K.

THE SPACE PROBE

The fundamental difficulties of telescopic observations of the planets from Earth throw into relief the importance of

---

*273°K (Kelvin) is the same as 0°C (Celsius or Centigrade). To get the more familiar °C from °K, simply subtract 273. The absolute zero of temperature (°K), "degree Kelvin" is −273 degrees below zero Centigrade or Celsius.

space techniques in advancing our understanding of the solar system. It is true that in the years following World War II the development of radio and radar astronomy led to important advances—the case of planetary distances and the problem of the rotation of Venus and Mercury have been mentioned. The study of the radio emissions from the Sun also led to great advances in our understanding of the solar atmosphere. Even so, there is no comparison with the revolutionary nature of the results achieved by the use of space vehicles in the twenty years since the launching of Sputnik 1 in October 1957.

The ability to place quite simple scientific instruments in orbit above the obscuring layers of the Earth's atmosphere immediately led to surprising discoveries. For example the Geiger counters in the first American Explorer satellite, early in 1958, revealed the zones of protons and electrons trapped in the Earth's magnetic field. This discovery led to the realization that interplanetary space was scarcely the near vacuum so far envisaged, populated only by the planetary bodies, the asteroids, meteors and comets. On the contrary interplanetary space was found to be a complex of particles and magnetic fields with the planets and Earth enveloped in the attenuated solar corona. Many ideas about the relationship of major geophysical effects, such as magnetic storms and the aurora borealis, to disturbances on the Sun needed revision.

The landing of Armstrong and Aldrin on the Moon in 1969 and their safe return to Earth with samples of lunar rock was a remarkable demonstration of the tremendous accomplishments of modern technology, but the public excitement tended to obscure the vital scientific features of the lunar expeditions. Only shortly before the landings distinguished astronomers were in dispute about the nature of the lunar surface—to the extent that some believed the surface to be covered with such a thick layer of dust that the astronauts would sink into it. Our ideas about the age of the Moon, the nature of the craters and the Moon's evolutionary history have been influenced by the American manned expeditions and the landings of unmanned lunar vehicles by the USSR.

But perhaps the most striking illustrations of the power of space techniques are the revolutionary discoveries made about the planets Venus and Mars. The dynamical problems surrounding Venus and the settlement of the rotation rate by radar have been mentioned. The wide disparity of views about the planet's rotation was matched by contrasting opinions about its physical condition. Extreme views were held—the surface was either a boiling ocean or an arid desert. There were those who took the natural view that the clouds must be substantially composed of water vapor, like the terrestrial clouds, with the possibility that the surface of the planet might be favorable for organic development. Some progress was made in 1932 when Walter S. Adams and Theodore Dunham at Mt. Wilson discovered the absorption bands of carbon dioxide in the spectrum of sunlight reflected from the atmosphere of Venus. In 1953 W. M. Sinton published new measurements of planetary temperatures and he obtained a value of -39 deg. C for the illuminated surface of Venus. This was in fair agreement with earlier estimates, and based on the information available in 1955 the two eminent astronomers, Donald H. Menzel and Fred L. Whipple[2] at Harvard, concluded that the surface must be entirely covered by ocean. At this stage several attempts were made, taking measurements from high-flying balloons, to find out if water or oxygen existed in the planet's atmosphere; but there was always the difficulty that the measurements had to be made through the Earth's atmosphere. As recently as 1964 in the English revision and translation of *The Flammarion Book of Astronomy* the comment was made about the clouds in the Venusian atmosphere that "To this day their composition is still one of the mysteries of the physics of Venus," and in the 1966 edition of the *Larousse Encyclopaedia of Astronomy,* after a careful review of the problem, the writer made the comment that "Our knowledge of the character of Venus' atmosphere is deplorably incomplete."

The first decisive indication that the ocean theory must be incorrect was obtained in 1956 when scientists at the Naval

Research Laboratory in Washington[3] succeeded in detecting the radio emission from the planet on a wavelength of 3.15 cm. They concluded that, contrary to Sinton's estimates, the temperature must be very high and they gave figures in the region of 560 to 620°K. The first attempts to send space probes to the planet were not successful—the USSR launched Venus 1 on 12 February 1961, and the USA, Mariner 1 on 22 July 1962—but on 27 August 1962 the United States launched Mariner 2 and this probe successfully transmitted data to Earth as it made a fly-by of the planet at a distance of 34,700 km. These data confirmed the belief that the surface of Venus must have a high temperature, and this revived the idea that, contrary to the Menzel/Whipple ocean theory, the surface of Venus must be a desert.

On 12 June 1967 the spacecraft Venera 4 was launched from the USSR. As this probe entered the atmosphere of the planet on 18 October of that year, a capsule of scientific instruments was released and a parachute system operated to slow the descent of this capsule through the atmosphere. As it descended, information about the constitution, pressure and temperature of the planetary atmosphere was transmitted to Earth—and there must be few examples in the history of astronomy where ninety minutes of observation have led to the resolution of so many contradictory theories. It is believed that the data transmissions ceased when the capsule was still 24 kilometers from the surface of the planet, but at that point the temperature was very high (500°K), the pressure was twenty times that of the atmospheric pressure on Earth and the atmosphere was over 90 percent carbon dioxide. Subsequent probes have been launched by the Soviet Union and the conditions at the surface of the planet are now known to be exceedingly hostile. The temperature is about 750°K, the atmosphere 97 percent carbon dioxide, and the pressure one hundred times that of the atmospheric pressure on Earth. Direct photographic evidence of these extreme conditions was provided by the Soviet spacecrafts Venera 9 and 10, which landed television cameras on the surface of Venus on

22 and 25 October 1975 respectively. Each probe transmitted for about sixty minutes, and the resulting pictures received on Earth are the first direct visual evidence ever obtained by man of the conditions on the surface of Venus. The pictures showed a large area of the landing site, which included scattered rocks with sharp outlines. Venera 10 landed 1,367 miles from Venera 9. The rocks at both sites were described as resembling "huge pancakes with sections of cooled lava or debris of weathered rock in between"—a landscape, according to Soviet scientists, typical of old mountain formations, and one which did not support the idea that the surface of Venus was a desert or an ocean.

There has been a similar discrepancy in astronomers' views of the physical nature of the planet Mars, even though, unlike the case of Venus, it has been possible to look at the surface of Mars through telescopes. As we have seen, turbulence in the Earth's atmosphere places a limit on the achievable resolution with optical telescopes of about 130 kilometers even at the closest approach of the planet to Earth.* And, since Mars is a bright object in the sky, relatively little advantage has been gained by using the large optical telescopes so essential for the study of the stellar Universe. Indeed, until the last few years, our information about the surface of the planet has rested on the sketches made by skilled observers viewing the planet directly through telescopes of moderate size. The first such sketch, revealing markings on the planet, is believed to have been made by Huyghens in 1659. This sketch and the more detailed ones, such as those made by Johann Schröter in 1800, all distinguished a dark area of the planet subsequently identified with the region known as Syrtis Major. Schröter believed this to be a cloud phenomenon in the atmosphere of the planet, and until recent years it was thought to be a deep depression on the planet's surface. Now it is recognized as a relatively smooth plateau distinguished primarily

---

*At the favorable 1956 close approach to Earth the planet appeared as a disk 25 seconds of arc in diameter—equivalent to a cent seen at a distance of 180 yards.

by its dark color. However, the greatest interest in Mars was stimulated by the drawings made in 1877 by Giovanni Schiaparelli, who used a 20-centimeter telescope for his observations. He identified many long intersecting straight lines on the planet, which he believed to be changing in size with time—the "canali." Then in 1895 Percival Lowell began an extensive study of these "canali" or "channels" from the Flagstaff Observatory in Arizona. He charted four hundred of them, some over 4,800 kilometers in length. Under the influence of Lowell the view became widespread that these were some form of irrigation system associated with the seasonal melting of the polar caps of the planet.

Primarily because of this work of Lowell and his belief, which he sustained against all opposition, that the markings he delineated were true canals of artificial origin, the planet Mars became a prime target of study and speculation in the search for extraterrestrial life forms. Indeed, there has been little scientific evidence even in contemporary times to make this search seem bizarre. At the 1956 close approach of the planet to Earth, William M. Sinton made spectroscopic observations using the 200-inch Palomar telescope and his results supported the view that vegetation might exist in some regions of the planet. Investigations of the constitution of the atmosphere suffered the same difficulties as in the case of Venus—namely, that the measurements inevitably had to be made through the atmosphere of the Earth. The balance of opinion was that nitrogen was probably a major constituent of an atmosphere much thinner than that of Earth—but there was no evidence of a poisonous condition or noxious gas, or of pressure or temperature such as exist on Venus. In fact, the noon surface temperatures at the equator of Mars were computed to be in the range of $268°K$ to $298°K$.

It is therefore easy to appreciate why Mars, like Venus, became an object of prime interest as soon as space technology made it possible to send probes to the planets and why the biological condition of the planet became a major aim of these investigations. The American Viking 1 and 2 spacecraft which landed on the surface of Mars on 20 July and 3 September 1976 were the first (and so far the only) probes from Earth to fulfill this ambition.* But before that time our knowledge of the planet had already been transformed by a succession of American Mariner spacecraft, beginning with Mariner 4 which made a fly-by of the planet at a distance of 9,800 kilometers on 15 July 1965. On 13 November 1971 Mariner 9 was placed in orbit around Mars, and by the time the spacecraft's transmitters were switched off nearly a year later, this probe had made 698 circuits of the planet and had transmitted to Earth 7,329 photographs of astonishing detail and clarity. The entire Martian surface was photographed to a resolution of one kilometer and large regions to a resolution of 100 meters—that is a thousand times better than the best of the photographs ever obtained by the earthbound telescopes.

The difficulty of determining the physical condition of the surface of Mars is well illustrated by this decade of close-up photography from space probes. For example, the regions of the surface photographed by the earlier Mariner fly-by probes were surprisingly heavily cratered. The photographs revealed a surface so similar to the lunar surface that the view that any water existed in the polar caps or elsewhere, or that conditions might be favorable for organic evolution, received a severe setback. But the total view of the planet given by Mariner 9 showed that, while there were indeed these heavily cratered regions, other regions of the planet were smooth—coincident with the classical "deserts" of the nineteenth-century sketches. The analysis of the Mariner pictures revealed mountain peaks rising to 26,000 feet, enormous cratered terrains twice the area of Texas, and canyons four times the depth of the Grand Canyon in Arizona. Perhaps most important of all, because of the stimulus which was given by the discovery to the Viking program, was the appearance near the

---

*Viking 1 was launched on 20 August 1975 and Viking 2 on 9 September 1975. A capsule from a Soviet space probe Mars 3 landed on the planet on 2 December 1971 but did not work successfully; neither did the Soviet Mars 4 and 5 in February 1974.

canyons of irregular linear features, closely resembling dried river beds. The conclusion was reached that these *were* dried river beds and that Mars now existed in an ice age with the water locked up in the polar caps or in subsurface permafrost.

The successful landing and operation on the surface of Mars of the two Viking spacecraft in mid-1976 must rank as one of the great triumphs of modern technology. The spacecraft operated continuously until 10 November 1976 when they were switched off because of a radio blackout occasioned by the conjunction of Mars and the Sun. Before the end of the year they were reactivated but already the programs originally planned for Viking had been completed. A long period will elapse before the vast amount of data can be analyzed and published.

Naturally, interest has been centered on the biological experiments included in the spacecraft. Each lander carried a trio of sophisticated experiments designed to search for biological activity in the soil scooped into the equipment by long-handled surface samplers under remote control from Earth. All these devices worked and gave positive results, but another device designed to identify organic molecules, and sensitive to a trace as minute as 0.05 parts in a billion, although working as intended, has given a negative result. These results have greatly puzzled the investigators and opinion is divided as to whether the positive responses from the three experiments are indicative of biological activity on the planet or are the results of some unexpected chemical reactions. One of the latest available comments from the director of Mission Planning and Scientific Analysis for Viking is that "The preponderance of scientific opinion is that most of what we have seen can be more easily explained by chemistry. However, it is by no means that clean cut. We need to do many more experiments with the Viking landers and on the Earth to answer this puzzle."[4]

By the end of 1976, it seemed that the results from Viking would not provide a categorical answer to the tantalizing question about the possibility that life may exist or has in the past existed on Mars. Even conclusive negative answers from Viking 1 and 2 might mean only that they landed in the wrong place, perhaps at the wrong season, or even that Martian life forms are entirely different from those on Earth.

A great mass of photographic, meteorological and atmospheric data has also been transmitted to Earth by the landers and orbiters of Viking 1 and 2. When the Martian atmosphere is clear, thousands of small rocks can be seen in the photographs taken by the landers and meandering features resembling dry river beds are easily identified. The temperature conditions on the surface can be compared with those of extremely severe arctic or mountainous conditions on Earth, extending from a low of about $-80°C$ during the Martian night to a high of about $-30°C$ during the Martian day. Estimates of the atmospheric pressure at the surface of Mars had been made on the basis of various optical observations from Earth and a value of about 65 millibars, about 16 times lower than the normal atmospheric pressure on Earth of 1013 millibars, had been assessed. The Viking measurements revealed that in fact the Martian atmosphere is very thin—7.7 millibars or 130 times lower pressure than that on Earth. Viking measurements to determine the constitution of the Martian atmosphere have also given surprising results. Twenty years ago the atmosphere was believed to consist largely of nitrogen. By 1970, after the analysis of early Mariner remote-sensing data, the opinion was that the atmosphere contained a considerable amount of carbon dioxide and carbon monoxide. Viking has shown that at the surface the atmosphere of Mars contains about 95 percent carbon dioxide, 2 or 3 percent nitrogen, 1 or 2 percent argon, less than one-half percent oxygen, and a very small amount of water vapor.

The information so far obtained from the Viking experiments has settled many questions about the present physical condition of Mars, but a host of new ones has arisen. Perhaps paramount are the apparent contradictory responses from the biological and organic molecular tests on the Martian soil. Another unsolved difficulty is related to the extreme thinness of the atmosphere. In the one to two percent of argon found in the Martian atmosphere, the Viking measurements reveal the ratio of the isotope argon-36 to argon-40 to be ten times lower than the ratio on Earth. The argon-36 isotope must be primordial, while argon-40 increases with time because of the radioactive decay of potassium-40. The difference in the ratio between the atmospheres of Earth and Mars may indicate that Mars has not outgassed to the same extent as Earth, and therefore that the planet has never possessed an atmosphere significantly denser than it does today. Scientists disagree about this—and on the vital question of whether the features resembling dried river beds once contained water. If they are extinct rivers, is the evidence from the Viking measurements sufficient to establish that the water is frozen in the polar caps or subsurface permafrost? And if so, will it ever be released again with an end of the present ice age on Mars, as happened on Earth several million years ago?

Venus and Mars were the obvious targets for the initial exploration of the planets by space probes—their orbits lie closest to that of the Earth's orbit around the Sun, and rocket technology made it possible to launch spacecraft which would reach these planets in a matter of months. The dispatch of spacecraft to the other planets presented a far more formidable problem. In the years following the initial successful flights to the vicinity of Venus (1962) and Mars (1964–65), American scientists developed the idea of using the gravitational pull of one planet to assist a spacecraft in its journey towards another, and the scheme of a Grand Tour, depending on planetary configurations in 1977 and 1979 which will not recur for another two centuries, evolved. Two major launchings were proposed, one in September 1977 and a second in November 1979. The 1977 space probe would pass Jupiter after 1.4 years, Saturn after 3 years and would then encounter Pluto after 8.5 years. The 1979 space probe would fly past Jupiter after 1.5 years, Uranus after another 4.2 years and would reach Neptune by the end of 1988. Unfortunately, the development of electronic equipment suitable for the journey would have been a formidable task. Such equipment would have to be capable of operating in space for long periods of time and would require nuclear power sources, as the Sun is too weak at the distance of the outer planets to provide solar power. With a cost estimate of about a billion dollars the projects were never initiated. Although the Grand Tour project never got beyond the planning stage, the concept of gravity-assisted flight has materialized and has already made it possible for the American spacecraft Mariner 10 to make close approaches to the inner planet Mercury, while in September 1979 the first encounter of a spacecraft with Saturn, after a gravity-assist from Jupiter, should occur. Since Mercury is never seen against a dark sky background, drawings made of the surface features, and indeed all the available evidence about the nature of the surface selected to make a "definitive" map of the planet, bore little resemblance to the photographs transmitted to Earth by Mariner 10.

Mariner 10 was launched on 3 November 1973 in such an orbit that it passed Venus at a distance of 5,750 kilometers on 5 February 1974. Assisted by the gravitational attraction of Venus the spacecraft then entered an orbit which brought it to within 750 kilometers of Mercury on 29 March 1974. Further close approaches occurred on 21 September 1974 and on 16 March 1975 and many of the photographs have resolutions of about 100 meters. A surface previously seen indistinctly is revealed as a mosaic of craters, scarps, ridges, circular basins and plains, bearing a strong resemblance to the lunar surface. It seems that Mercury may have had an evolutionary history similar to that of the Moon—a heavy bombardment in the early stages followed by widespread volcanic activity. It has only a tenuous atmosphere, probably consisting mainly of helium. A surprising feature revealed by Mariner 10 was that Mercury has an intrinsic magnetic field.

Two American spacecraft, Pioneer 10 and 11, have already transmitted data during a successful fly-by of the planet Jupiter. After a flight of 21 months Pioneer 10 passed within 130,400 kilometers of Jupiter on 4 December 1973. The orbit was planned so that in the gravitational interaction with Jupiter, the spacecraft was accelerated and will eventually move out of the solar system. A year later on 2 December 1974 Pioneer 11 passed 26,000 miles above the cloud tops of the planet; it was within a million miles of Jupiter for less than 48 hours.

There were many remarkable features of these two brief encounters with Jupiter, perhaps the most notable being the detailed photographs of the enigmatic great red spot, now generally thought to be the vortex of a violent and long-lasting cyclonic storm in the Jovian atmosphere extending about five miles higher than the surrounding cloud tops. Radiometer measurements gave the surprising result that Jupiter is radiating two or three times as much energy as it receives from the Sun, energy perhaps derived from the continuing gravitational contraction of the planet. The measurements of the Jovian magnetic field, of the charged particles in the zones of trapped radiation around the planet, and indeed the very fact that the instruments in the Pioneers survived the transit through these zones, have provided a wealth of data about the planet.

The orbit of Pioneer 11 in the vicinity of Jupiter was chosen so that the gravitational interaction would swing the spacecraft into a new orbit, which after five years and another one and a half billion miles of travel will bring it in September 1979 to the vicinity of Saturn. According to present plans, the orbit will be adjusted by thrusters to take it between the globe of the planet and the innermost ring. It is hoped that by passing within 2,300 miles of the cloud tops of Saturn the probe will avoid damage from the fragments of material in the rings. When it eventually leaves Saturn, Pioneer 11 will pass about 12,000 miles from Titan, the largest of the planet's satellites.

THE FUTURE INVESTIGATION OF THE SOLAR SYSTEM

Speculation about the trend of future investigations in astronomy is hazardous in an epoch of revolutionary discoveries, but it seems evident that the future in this case belongs to the technique of the space probe. Almost every investigation of the planetary system which has been carried out by telescopes on Earth can now be performed more concisely and perfectly by instruments carried to the vicinity of the planets. The question of the future therefore resolves itself into the time spans over which the type of space probe observations already made on Mars, Venus, Mercury and Jupiter can be extended to the remaining planets and possibly to some of the larger satellites of Jupiter. In some cases the technical ability already exists, and the question is one of priorities within the space programs.

The next logical step in the investigation of Mercury would be to place a probe in orbit around the planet—a mission already tentatively scheduled in the NASA program for the next ten to fifteen years. Within that time span it also seems possible to follow the Viking probes to Mars with a probe which will not only scoop surface soil into a lander as Viking has done, but will also return it to Earth, as the Soviet probes have done with lunar samples. The return of planetary samples to Earth from Venus can also be envisaged, but in this case the operation seems more likely to follow a detailed investigation of the atmosphere of the planet by probes which can remain at various heights in the atmosphere long enough to determine the planet's environment in more detail.

The technique of gravitational assistance has eased the problem of the rocket power required to reach the outer planets and there seems to be no technical reason why close approaches to all the major planets, with orbiters in the case of Jupiter and Saturn, should not be realized by about 1990. It will be surprising if close fly-bys of the larger asteroids and of the satellites of Saturn and Jupiter have not also been achieved by then. In addition, in 1986 Halley's comet is at perihelion and a close fly-by to obtain photographic and other details of the nucleus and tail of the comet is an important scientific requirement.

The question of manned flights to the planets raises problems of a different order of magnitude because we are concerned in these cases not merely with technological ability but with human physiological and psychological problems. In the lunar case manned flights followed the automatic explorations with remarkable speed—the entire US program was encompassed within the 1960–70 decade. However, before the first journey to the Moon men had already survived in Earth orbit for periods longer than the lunar journey required.

US scientists have proposed a similar manned program for Mars; it was argued that if sufficient money and priority could be given to the project a manned expedition to Mars would be feasible in the period 1980–90. Though Soviet intentions remain unknown, at least for the USA no such expedition appears to be in prospect for a long time ahead. The space shuttle operations now envisaged as a technique both for the USSR and the USA will undoubtedly place these planetary missions in a new perspective and the possible time spans for all types of planetary explorations may be affected significantly by experiences with the shuttle. At least one thing seems beyond contention. If space probe flights into the planetary system continue even at the level of the last few years, then our knowledge of the system is bound to suffer further revolutions as details of the major planets are revealed. The stage at which the succession of discoveries will lead to an unambiguous view of the formation and evolution of the system cannot be judged.

Although it is possible to be optimistic about the progress of these investigations, there must be greater reserve about the possibility of resolving the vital question of biological activity elsewhere in the solar system. The complexities of the contemporary Viking investigations, which have given neither a positive nor negative answer, illustrate the very great difficulties of remote sensing. Neither is time and technological advance on the side of the biological investigations, since the dangers of contamination by terrestrial organisms will increase with each flight and will reach a critical stage when man himself travels to a planetary body.

IV

The Doctrine of Many
Possible Worlds

The doctrine of many possible worlds was enunciated by
Leibniz in the late seventeenth and early eighteenth centuries. According to Leibniz there was an infinite number of
possible worlds, all contemplated by God before He created
the world on which man lives. He argued that God created
this world because it was the best of all possible worlds, possessing the greatest excess of good over evil.

Leibniz died in 1716 and more than two centuries elapsed
before the concept of a multiplicity of worlds became more
than a matter of philosophical contemplation. Then a distinguished astronomer—Harlow Shapley, the director of the
Harvard Observatory—made an astronomical speculation.
He concluded that in the entire Universe there must be some
10^{20} stars with characteristics like those of the Sun. He
thought that a reasonable assumption would be to consider
that one in a thousand of these stars would have a planetary
system, that one in a thousand of these would have a planet
at the right distance from the star to give temperature conditions similar to those on Earth, that one in a thousand of these
would have the right mass and size to retain an appropriate
atmosphere and so on. His eventual conclusion was that there
were a billion stars in the Universe with a planet on which
conditions could, in principle, favor the emergence of life.

The nebular theory of planetary evolution, coupled with

the modern view that the stars themselves are formed by condensation from clouds of dust and gas, makes it possible to seek a more scientific view of the number of stars which may have a planet suitable for organic evolution. There are some 100 billion stars in the Milky Way system spread over an enormous range of masses, temperatures and luminosities. The Sun seems to have average properties among the stars. For example, the luminosities (that is, the total power radiated by a star) cover a range from a millionth to one hundred thousand times the luminosity of the Sun. The surface temperature of the Sun is 5,800°K, whereas the temperatures of other stars range from 2,000 to over 60,000°K, and the masses from a few percent to over one hundred times the mass of the Sun. The developments in theoretical astrophysics during the last twenty years have led to a fairly comprehensive understanding of the evolutionary history of stars over this great range of parameters. In the contraction from the initial globule the temperature and pressure in the interior rise until thermonuclear reactions involving the conversion of hydrogen to helium commence. The contraction ceases and a long equilibrium is established—a star in this condition is said to be on the main sequence. The eventual lifetime in this stable condition for a star like the Sun is about 12 billion years; then, when the hydrogen supply begins to fail, the star evolves away from the main sequence and its subsequent fate depends critically on its mass.

At this stage the effect on any planetary system would be catastrophic, so that clearly as far as organic evolution is concerned, any interest must be concentrated on the stars with temperatures not too dissimilar from the Sun's and which have been established on the main sequence for several billion years and have a similar future life expectancy. Those who have considered this problem take the view that the stars on the main sequence in the spectral class range F to K (the Sun is spectral class G) with temperatures in the range from about 4,500 to 7,500°K are the optimum candidates. These include about a quarter of all the stars in the Milky Way. Even making allowances for binary and multiple star systems and others which may be excluded for various reasons, this leads to an estimate of about 20 billion stars in the galaxy which may have a planet with Earth-like characteristics.

Within the field of view of the largest optical telescopes there are about 100 million galaxies. If all of these were spiral galaxies with characteristics similar to the Milky Way then similar considerations would imply that there would be 2×10^{18} stars in the observable universe which might have planetary systems capable of supporting organic evolution. Of course, the estimate is fallacious. From the characteristics of the extragalactic nebulae we know that a large percentage are undergoing violent evolutionary processes; even so, Shapley's estimate of a billion stars in the Universe with a planet which might favor the emergence of life is seen to be very cautious. The astronomical answer from these theoretical considerations seems inescapable: there are billions of stars even in the observable Universe which could provide a suitable environment for a habitable planet.

The weakness of these arguments is that no planetary system has ever been observed around even the stars nearest to the Sun, nor would one expect to be able to make such an observation in view of the poor resolving power of telescopes on Earth coupled with the huge difference in brightness between star and planet. However, a certain amount of circumstantial evidence does exist for the presence of planets around some of the nearby stars. The most compelling is undoubtedly that resulting from the observations of Peter Van de Kamp of Swarthmore College. He studied the motion of one of the nearest stars to the Sun—known as Barnard's star—for twenty-five years and in 1963 he announced that there were irregularities in its motion across the stellar background which could be explained if there was a planet about the size of Jupiter in orbit around the star. Subsequently he concluded that the irregularities in the star's motion implied the existence of a second planet and in 1971 Graham Suffolk and David Black at the Ames Research Center of NASA analyzed again the whole of Van de Kamp's data and concluded that the observations could be explained most satisfactorily if three planets were in orbit around the star. Barnard's star is a red dwarf with a mass only about 15 percent of that of the Sun. Of course it is only with stars of small mass that one could expect to detect irregularities in motion occasioned by a planet of the size of Jupiter. About eight other dwarf stars, within about twenty light years of the Sun, are found to have similar irregularities in motion compatible with the presence of an unseen companion having a mass of 10 to 35 percent of that of the star.

Today there are responsible sections of the scientific community who do not question that elsewhere in the Universe there are planetary systems on which intelligent life has developed. Furthermore, attempts are being made to establish communication with extraterrestrial communities. The rapid contemporary growth of interest was initiated by the publication in *Nature* on 19 September 1959 of an article by two scientists from Cornell University, Giuseppe Cocconi and Philip Morrison.[1] Their thesis was that no theories existed which enabled a reliable estimate to be made of the probabilities of planet formation, of the origin of life or of the evolution of societies possessing advanced scientific capabilities. In the absence of such theories they suggested that our own environment implied that stars like the Sun with lifetimes of billions of years would possess at least one planet on which life had developed. Further, their view was that among such systems there might be some societies which had maintained themselves for a very long time compared with the time scale of human history, and that these might have scientific interests and techniques much greater than those available to us. According to Cocconi and Morrison, such an advanced society would have recognized our own planetary system as a probable place for life to develop and would long ago have transmitted signals to us in the hope of eliciting a response. Furthermore, Cocconi and Morrison suggested that the communication would be carried out on an objective standard of frequency throughout the whole Universe: the spectral radio emission line from neutral hydrogen on a wavelength of 21 centimeters. Their article ended with a plea that attempts should be made to detect such signals emanating from some of the nearer stars, which they specified.

In retrospect it can be seen that this article was published at a most appropriate time. First, the authors had calculated that the sensitivities of the radio equipment and radio telescopes now available were adequate to detect signals from a community living in the vicinity of the nearer stars—making reasonable assumptions about their transmitting facilities. Second, contemporary opinion favors the nebular hypothesis for the development of the solar system. The older catastrophic theories, involving processes whereby another star interacted with the Sun, supported the concept that our planetary system was unique, or at least a rare occurrence in the Universe. Third, certain observational developments have stimulated scientific interest in the possible existence of extraterrestrial life-forms: surprisingly, a number of organic molecules have been discovered in interstellar space.

ORGANIC MOLECULES IN SPACE

The photographic evidence for the initial globules and protostars, which strongly supports the contemporary view that the stars condensed from the interstellar gas clouds, comes from the luminous clouds of hydrogen gas in the neighborhood of hot stars—for example, the Orion nebula. In these regions the radiation from the stars ionizes the hydrogen gas. The subsequent energy transitions of captured electrons emit light, and the cloud of hydrogen is seen as an emission nebula. Until the development of radio astronomy after World War II these emission nebulae accounted for almost all of our observational knowledge of the interstellar medium, although from considerations of the dynamical condition of the Milky Way it was presumed that much more gas and dust must exist unseen in space than that revealed in the emission nebulae. In 1951 observational evidence in support of this belief was obtained when, almost simultaneously, observers in Holland, the USA and Australia announced that they had succeeded in detecting the radio spectral line on a wavelength of 21 centimeters from neutral hydrogen in the Milky Way. The spectral line—generated by a reversal of the electron spin in the magnetic field of the proton in the hydrogen atom—is emitted on a precise frequency, but the frequency of the emission observed by the radio telescopes is shifted because of the motion of the clouds of neutral hydrogen relative to the Earth. A survey of these frequency shifts soon made it possible to delineate the motions of the galaxy in regions which were obscured from the view of optical astronomers because of the dust clouds in space.

In a few years the first observational picture of the detailed spiral structure of the Milky Way was obtained. Within this detailed structure the distribution of the hydrogen gas could be measured. There are extremely wide variations. For example, within a radius of two thousand light years of the galactic center the total mass is equivalent to that of a billion suns, but nearly all of this is in the stars—only about 1 percent is in the form of gas. On the other hand, in the spiral arms in the general vicinity of the Sun the situation is totally different. Here gas accounts for about a fifth of the total mass. It is in these regions of the Milky Way, which contain the great gas clouds from which stars are forming, that later discoveries have been of extreme importance in relation to the possibility of organic development.

The belief that in the spiral arms the 20 percent of interstellar material was entirely hydrogen gas—with a dust content of about 1 percent of the mass of the gas—did not survive for long the probings of the radio telescopes using new techniques with which the spectral line from neutral hydrogen had been detected. In 1963 the spectral line on a wavelength of 18 centimeters, characteristic of energy-level transitions in the hydroxyl molecule (OH), was discovered in the interstellar gas clouds.

During this period great progress was being made with the development of radio receiving techniques and the construction of accurately figured radio telescopes for observations on much shorter wavelengths, in the millimeter range of the spectrum. These new facilities, equipped with advanced digital search and data handling techniques, have led to a rich harvest of discoveries of spectral lines characteristic of other molecules in the interstellar gas clouds. In 1969 the 6 centimeter spectral line characteristic of the formaldehyde molecule (H_2CO) was discovered. Three years later the spectral lines of twenty-five molecules had been discovered, and today the existence of forty different kinds of molecules in the interstellar medium has been established. And it may be of great significance that the water molecule (H_2O) is included in this list; indeed, if our eyes were tuned to a wavelength of 1.35 centimeters the emissions from the water molecules would be the most prominent objects in the sky.

The complexity of other molecules in the interstellar clouds is accepted by many scientists today as evidence that the basic material for organic evolution exists in space—and in the regions where star formation is occurring. Although these molecular constituents make only a small addition* to the mass of the interstellar hydrogen gas, their discovery may well be of great significance in attempts to understand the processes of organic evolution in the Universe.

THE EVOLUTION OF HABITABLE ATMOSPHERES

Theory and observation have brought us to a remarkable conclusion. Within the universe observable with large contemporary telescopes there may be billions of planetary sys-

---

*This comment should not be allowed to obscure the fact that vast amounts of this molecular material are distributed throughout the gas clouds. In a paper published in the *Astrophysical Journal* (196, L99 1975) a group of American astronomers described their successful search for the spectral line characteristic of ethyl alcohol in the Sagittarius interstellar cloud. They calculated that in this cloud there was enough alcohol to make 10^{28} bottles of whiskey—more alcohol than man has made in all recorded history.

tems which are capable of providing a suitable environment for habitable planets. The discovery of complex molecules in the gas clouds from which these stars condense seems to be a powerful indicator that life may have evolved elsewhere in the Universe. But apart from these two remarkable circumstances there are other vital questions which must be considered: will the right kind of atmosphere develop on other planets and, if so, will life emerge inevitably in that environment?

The formation of stars and planetary systems on the nebular hypothesis implies that the planets will be formed with approximately the same composition as the star. In the case of the Sun this means about 2 percent heavy elements and the remainder hydrogen and helium. Indeed the outer planets do have approximately this composition. Earth clearly does not, since we should be unable to live in such a hydrogen-helium environment; in fact the fraction (by weight) of helium in our atmosphere is 0.72 millionth and the fraction of hydrogen is only 0.04 millionth. Hence the Earth was either formed without helium and hydrogen in the atmosphere or lost these gases soon after formation. A planet will not retain hydrogen or helium unless it is massive enough to prevent the escape of the gaseous molecules. Computations indicate that a planet like Earth could retain a hydrogen atmosphere for only about one thousand years, and helium for about 10 million years. Hence there is no difficulty in understanding why Jupiter has retained hydrogen and helium, unlike Earth, Venus or Mars. The problem is that Earth has an atmosphere differing markedly from that of the primeval nebula in other ways. For example, argon and neon should be retained in the Earth's atmosphere for an immense time (10^{200} years), yet the actual proportions of these gases in the atmosphere are far below those to be expected from their relative cosmic abundance compared with other heavy elements on Earth. Furthermore, after helium, oxygen, nitrogen and carbon are the next most abundant elements cosmically, and likewise presumably in the nebula from which the Sun condensed. Carbon readily combines with hydrogen to form methane, which must have been a major constituent of the primitive atmosphere. Since oxygen and nitrogen combine with hydrogen to form water vapor and ammonia, these gases must also have been abundant in the early history of the Earth, but the fraction by weight of water in the present atmosphere is only 0.017 and there are no measurable amounts of ammonia. The free oxygen would probably combine with iron and silica in the crust of the Earth to form solid compounds.

Indeed, the present atmosphere of Earth differs so markedly from any conceivable primeval atmosphere that an inevitable conclusion seems to be that at some stage the early atmosphere disappeared in some cataclysmic event. No conceivable terrestrial reactions could account for the present scarcity of the inert gases argon and neon. The forces that stripped away the primitive atmosphere may have been associated with a violent solar eruption, or it could be that the inner planets were raised to a high temperature by bombardment with debris in the early formative stages. In any case, it seems that our present atmosphere is a secondary event and a widely accepted opinion today is that the gases of which it is composed were subsequently exhaled from the solid body of the Earth. For example, if volcanoes have been as frequent in the past as they are today then there would be no problem in accounting for the amount of water on Earth. Volcanic activity alone through the long history of the Earth would provide enough water vapor to fill all the oceans. Similarly volcanic eruptions could have supplied all the nitrogen which we find in the present atmosphere.

The volcanic gases also contain 10 percent carbon dioxide. Where is this gas? If the theory is correct, the concentration of carbon dioxide in the atmosphere should be much greater than the three-hundredth percent of the present atmosphere. It is believed that most of this carbon dioxide was dissolved in the oceans of the Earth, a process leading to the formation of carbonates, and was taken up in reactions with silicates in the crust.[2] When living organisms emerged on Earth the removal of carbon dioxide was further assisted. The growth of algae, plankton and vegetation played a further vital role in absorbing the carbon dioxide and by photosynthesis returning the residual oxygen to the atmosphere. As fossil hydrocarbons such as coal formed, this carbon was locked in the Earth's crust, rather than being returned to the atmosphere as carbon dioxide following oxidation of dead organic matter. Furthermore, were it not for this process of oxygenation we should not find oxygen in the Earth's atmosphere because it is so reactive that it would be depleted by chemical reactions with the Earth's crust.

Clearly the emergence of the Earth's habitable atmosphere has involved a series of delicate balances throughout a few billions of years—and in these balances the emergence of primeval organisms and plant life has been a critical factor. The delicacy of the balance is even more manifest when we compare Earth with the two neighboring planets, Venus and Mars. Venus is so similar to the Earth in mass and size that in the early stages the two planets must have been almost identical. For the first billion years at least, they would have generated the same internal heat through radioactive processes and must have suffered the same level of volcanic activity. Indeed, the huge amount of carbon dioxide now known to exist in the atmosphere of Venus is about that to be expected from the accumulation of volcanic gases. But the processes which reduced the carbon dioxide level on Earth did not occur on Venus, where the accumulation of this gas trapped the solar radiation (the greenhouse effect) and the surface temperature of the planet increased to the present very high level. The reason is believed to be connected with the fact that Venus moves in an orbit closer to the Sun than Earth:—the solar flux is twice as great. The temperature on the surface of the planet always remained above the boiling point of water at the pressures involved, so there were no oceans to dissolve the carbon dioxide as happened on Earth. The atmospheric-crust reactions are also temperature dependent, and at the higher temperatures on Venus large quantities of carbon dioxide could accumulate in equilibrium with the silicates.[3] The equilibrium between the production of carbon dioxide on Earth and its depletion by absorption in the oceans and by reactions with the Earth's crust was never established on Venus, and the accumulation of this gas and the increasing temperature probably militated against the emergence of even primitive organisms on the planet.

In the case of Mars, opposite circumstances have prevailed. The smaller size of the planet and its greater distance from the Sun have diminished the processes of internal heating and volcanic activity. The atmosphere is very thin and the surface of the planet is cold. Whether these circumstances have entirely prevented the emergence of living organisms and plant life will not be known until the Viking landers and their successors yield definitive answers.

From the study of the planets Earth, Venus and Mars it seems that the conditions for the evolution of a habitable atmosphere depend critically on the mass of the planet and its distance from the Sun. We do not know what the limits are, even for our own solar system. The physical similarity of Earth and Venus and their radically different atmospheres demonstrate that a planet with the size and mass of the Earth could not support life if it were much closer to the Sun than is the Earth. On the other hand, if Mars had the size and mass of Earth or Venus it is possible that a habitable atmosphere could have developed, although at that distance the solar flux would be less than half that on Earth.

For planetary systems around stars whose luminosity differs from the Sun, another important variable has to be considered. Some authorities believe that the distribution of planetary distances around stars of differing luminosity may be in a similar ratio to that found for the Sun, and that there will be the same probability of finding planets within the appropriate distance from the star yielding the correct temperature for the evolution of a habitable atmosphere. As regards the range of stellar luminosities to be considered, the stable lifetime of a star on the main sequence is the critical factor and this excludes stars hotter than about 7,500°K, while with dwarf stars problems of tidal interaction between planet and star arise.

The weakness in all these considerations is lack of observational evidence. Even a little knowledge of conditions in another planetary system would be of inestimable significance. As it is, the discussion of the possible number of habitable planets in the Universe reduces to the assumption of mediocrity—that half of all planetary systems should have at least one favorably situated planet.

THE EMERGENCE OF LIVING ORGANISMS

If planets exist elsewhere in the Universe with habitable atmospheres there remains the vital question about the emergence of living organisms. If the appropriate atmosphere exists do living organisms arise as an inevitable consequence? Indeed, the development of the habitable atmosphere on Earth has been essentially related to the emergence of living organisms because of their critical rôle in the absorption of carbon dioxide. Once more the only available evidence about living organisms exists on Earth. This evidence implies that for about the first billion years the Earth was barren, then living organisms appeared and after another three billion years biological evolution reached the stage which we witness today.

Do we believe, like Aristotle, that living beings exist as they are because God made them thus and that the species of life have existed for infinite time? Or do we share the view of Kant, inherent in his *Critique of Judgement,* that God used natural laws of growth and transformation in creating living beings—thereby reserving the action of divine reason within the mechanical processes of science? Or do we believe like Henri Bergson in the disparate worlds of life on the one hand and inert matter on the other and that the Universe is the clash and conflict of these motions:

The animal takes its stand on the plant, man bestrides animality, and the whole of humanity, in space and in time, is one immense army galloping beside and before and behind each of us in an overwhelming charge able to beat down every resistance, to clear many obstacles, perhaps even death.

During the last century the possible extremes of philosophical attitudes have been narrowed by astronomical and biological discoveries.

The prejudices of thousands of years were swept aside by the work of Louis Pasteur. The prize awarded to him by the French Academy of Sciences in 1862 recognized the conclusive nature of his demonstration that microorganisms were not spontaneously generated by various infusions and solutions of organic substances, but were present in the air and on the surface of common objects around us. He destroyed the belief in spontaneous generation so long held to be an incontrovertible and self-evident truth.

In the late nineteenth century, however, the work of Pasteur appears to have led to a false trail in the search for the origins of life, for his experiments were regarded by many scientists as conclusive evidence that the metamorphosis of inorganic materials into living organisms was impossible. In his Presidential Address to the British Association in 1871 Lord Kelvin said, "Dead matter cannot become living without coming under the influence of matter previously alive. This seems to be as sure a teaching of science as the law of gravitation." Such attitudes exemplified the idealistic concept that, since spontaneous generation was impossible, therefore life must be eternal.

Scientists then had to face the problem of how life arose on Earth and this encouraged the revival of the ancient doctrine of panspermia, a belief that the germs of life are dispersed everywhere throughout the Universe. At the beginning of this century there was much discussion of the means by which life could have been transported to Earth from elsewhere in the

Universe. There were many adherents of the theory of lithopanspermia, according to which the germs of life or perhaps even complex living organisms were brought to Earth by meteorites or cosmic dust. Then in 1906 the Swedish physical chemist Svante Arrhenius introduced the revolutionary idea that the transportation of living organisms from one heavenly body to another did not necessarily require any material carrier, but that the pressure of light was the main agent. The fact that light rays did exert a pressure had recently been demonstrated experimentally and Arrhenius calculated that bacterial spores with diameters of about 2×10^{-4} millimeters could travel through space at great speed under the pressure of sunlight or starlight. Scientific support for the radiopanspermic theory of Arrhenius continued in various forms for more than twenty years, but about fifty years ago it was demonstrated that living organisms could not survive the intense ultraviolet radiation to which they would be subjected in their protracted journey through space—Arrhenius had calculated that a spore would need nine thousand years to travel the distance separating the Sun from the nearest star.

Throughout this period there were distinguished scientists, following the work of Pasteur, who could not reconcile themselves to this train of development. For example, in their Presidential Addresses to the British Association in 1870 and 1874 both T. H. Huxley and J. Tyndall promulgated their opinion that life originated from lifeless matter. Kelvin's opinion quoted above was, in fact, an attempt to rebut the view of Huxley, who in his 1870 address said, "If it was given to me to look beyond the abyss of geologically recorded time . . . I should expect to be a witness of the evolution of living protoplasm from not living matter." Although Charles Darwin was primarily concerned with the evolutionary principles leading to the development of higher organisms from lower ones he also was in sympathy with the idea that life developed from non-living matter.

The real beginning of the scientific revolution which led to the contemporary views on the origin of life began in 1922 when the Soviet scientist A. I. Oparin suggested that the first living things to develop on Earth were able to nourish themselves on organic substances which had been formed on Earth long before the appearance of life.[4] Oparin believed that it was first necessary to "Clarify our ideas on the following question: What were the natural conditions during the formation of the Earth or in the early stages of its existence which led to the emergence of the hydrocarbons and their simplest derivatives? For these are the carbon compounds from which there could later arise all those other extraordinarily complicated organic substances which form the material basis of life." A few years later in 1929 J. B. S. Haldane also proposed that the development of organic compounds took place before the formation of the first living organisms.[5] Oparin and his followers developed in the succeeding years a complete theory of the processes by which complex organic compounds, making use of available energy sources either from the solar radiation or from electrical discharges in the atmosphere, might have developed from the gases in the primeval atmosphere of the Earth. The first demonstration that such processes could occur under terrestrial conditions was given in 1952–53, by a student of Harold Urey. At his suggestion, S. Miller mixed together the gases hydrogen, ammonia, methane and water vapor, all of which are believed to have been present in the early atmosphere of the Earth, and circulated the mixture through an electrical discharge. After a week he found that the water contained several types of amino acids. Other complex organic molecules, including nucleotides, have since been produced from the mixture of gases believed to have been present in the early atmosphere of the Earth. Furthermore, various forms of energy including ultraviolet radiation, heat, and bombardment by elementary particles as well as electrical discharges have been effective energy sources in the synthesis.

As organic molecules were formed in the early atmosphere they were probably dissolved in the oceans and over a long period of time a dilute concentration of amino acids and nucleotides accumulated in the primeval seas of the Earth. The process must have continued for millions of years. In the light of contemporary evidence there seems to be adequate support for Oparin's view that organic material was formed in this manner on Earth long before living cells emerged.

In 1967 E. S. Barghoorn and J. W. Schopf published the results of their investigation of an ancient rock formation in South Africa in which they found microstructures which could be the fossils of single-celled organisms. The rocks are 3.2 billion years old. If these are the fossils of living cells, then life must have emerged on Earth about 3.5 to 4 billion years ago. How did this critical transition occur—when the organic molecules in the early oceans formed into very large complex molecules, to give the proteins, enzymes and nucleic acids now known to be the crucial molecular constituents of living organisms? No one is yet certain, but many scientists now engaged on the problem believe that an explanation will be found.

The problem is immensely complex. The amino acids and the nucleotides, believed to have been formed from the primeval gases, are themselves complicated organic molecules, each containing about thirty atoms of hydrogen, nitrogen, oxygen and carbon. As a result of the brilliant work of the molecular biologists during the last twenty years it is now established that about twenty different kinds of amino acids and five different nucleotides play a critical rôle in living organisms. Within the living cell the amino acids are linked to form proteins—one class of protein forms the structural material of the cell and another class (the enzymes) controls the chemical reactions necessary to sustain the life of the cell. The proteins are assembled in the living cell by the nucleotides, which are joined together in the long chains called nucleic acids. The classic paper of Watson and Crick,[6] which appeared in *Nature* on 25 April 1953, elucidated the structure and vital function of one of these nucleic acids, DNA (deoxyribonucleic acid). It is the genetic substance which controls the assembly of proteins in the cell determining the different forms of life. In advanced organisms the DNA molecule contains 10 billion atoms.

These modern discoveries have illuminated the processes of evolution from the simplest living organism. The remarkable ability of this DNA molecule to copy itself is the essential factor determining inheritance. Once the DNA molecule was formed, the cell could divide, each new cell possessing a copy of the DNA molecule and thus resembling the parent. Evolution occurs through random alterations of the DNA strand—and only those changes which give rise to beneficial interactions of the organism with the physical environment enable the organisms to survive. Thus the process of inheritance and the scientific basis underlying the phenomenon of natural selection described by Darwin in 1859 are explicable. Yet, while most scientists would now agree that the description first given by Oparin of the formation of organic molecules in the early seas from the gases of the primeval atmosphere is correct, and while they would also agree that the basic functions of the proteins and nucleic acids, and the vital part played by the DNA molecule in determining species, inheritance and evolution are understood, there is still a most vital gap in our knowledge. We do not yet have a conclusive answer to the question raised above as to the processes by which the molecules in the early seas formed themselves 3 or 4 billion years ago into the complex molecules to make the transition from non-living to living material.

Today the study of the question of the origin of life on Earth is centered critically on this epoch. Before that time the macromolecules which were present in the ocean were formed by a synthetic process using the gases in the primeval atmosphere, and this process can be demonstrated in the laboratory without difficulty. After that time—that is, after the first single-celled organisms evolved—there is a combined geological and biological explanation of the evolutionary processes which have led to the existence of man today. However, our knowledge of the processes by which the transition was made during that epoch of a half billion years, just over 3 billion years ago, is speculative and fragmentary. There are many who believe that it is futile to search for life elsewhere in the Universe, or even to consider the issue, until we can solve this problem of the origin of life within the boundaries of our own planet.

THE IMPROBABILITY OF EXISTENCE

The assessment of the probability that life has emerged elsewhere in the Universe is at present faced with dilemmas which have both astronomical and biological origins. There is first the assessment of the probable number of habitable planets in the Universe. The belief in a multiplicity of planets has a theoretical basis but is supported only by very tenuous circumstantial observational evidence. Even given the multiplicity of planets it seems that the growth of habitable atmospheres requires a most delicate balance of temperature and gravitational conditions. The space probe measurements on Mars and Venus serve to illuminate the delicacy of the terrestrial situation. If in spite of these uncertainties we assume that because of the great numbers of stars in the Universe there are other planets besides Earth with suitable environments, then the biological problems must be faced.

The difficulties of reaching any understanding of the complex processes which occurred more than 3 billion years ago are emphasised when the extreme improbability of the emergence of the cell from the amino acids and nucleotides of the primeval seas is considered. With our present knowledge of the number of different amino acids, nucleotides and chain lengths in living organisms, it can be calculated that there are some 10^{130} possible alternative sequences. This is a measure of the complexity of one of the smallest enzymes. The smallest units possessing their own metabolism—for example single cells such as the bacterium *Escherichia coli*—contain one large nucleic acid molecule comprising three to four million base pairs arranged consecutively in a double-stranded helical structure. The gene complement of this bacterium offers some $10^{2,000,000}$ alternatives. These numbers are incomprehensible even in terms of astronomical quantities—the total number of hydrogen atoms in the observable universe is only of the order 10^{78}. The length of the *E. coli* bacterium is about 2 microns and its mass 5×10^{-13} grams. If the bacterium is placed in a small quantity of water containing a few milligrams of glucose and some mineral salts (nitrogen, phosphorus, sulphur, etc.), then within 36 hours this solution will contain several million bacteria. The problem of origins is therefore, how did this critical selection of this combination of molecules occur out of the gigantic range of possibilities? The only generalized scientific answer which can be given today is that over the long time period of a half billion to a billion years, the processes of chance and natural law operated in the primeval seas—chance, in that vast numbers of differing assemblies of the molecular constituents took place, and natural law in that only the fittest survived according to the laws of natural selection appropriate to the boundary conditions of that environment.

This generalized proposition—that processes of chance *and* natural law led to the emergence of living organisms from the relatively simple organic molecules in the primeval seas—is valid only if the probability of the right assembly of molecules occurring is finite within the time scale envisaged. Here there is another great problem. In the example already given of a relatively small protein molecule with 100 amino acid residues, selection of *this* sequence of residues had to be made *by chance* from 10^{130} alternative choices. The operation of pure chance would mean that within the half billion to a billion year period the organic molecules in the primeval seas might have to undergo 10^{130} trial assemblies in order to hit upon the correct sequence. The probability of such a *chance* occurrence leading to the formation of one of the smallest protein molecules is unimaginably small. Within the boundary conditions of time and space which we are considering it is effectively zero.

Nevertheless, the presence of ourselves on Earth today is

evidence that a sequence of similar events of almost zero probability did take place over 3 billion years ago. Within the last ten years scientists have made some progress in the attempt to understand the conditions which may have operated to raise the probability to a finite level. For example, the particles known as ribosomes are essential components of the mechanism that translates the genetic code. They have molecular weights of about a million and are an assembly of about thirty different proteins and three different nucleic acids. It has been demonstrated that the dissociated constituents of the ribosomes will spontaneously reassemble themselves into particles precisely the same as the original material, that is, with the same composition, molecular weight and functional activity as the original. In 1966 American scientists R. S. Edgar and W. B. Wood published details of their work on a bacteriophage (a virus which attacks bacteria) known as T4.[7] The bacteriophage has a complex and precise structure. It also operates as a piece of microscopic precision machinery since it functions by attaching itself to the host cell and injecting into the cell its DNA content. In this experiment, different parts of the bacteriophage were obtained separately from various mutants of the virus, and yet when mixed together they spontaneously reassembled themselves into bacteriophages identical in structure and function to normal ones and capable of exercising the DNA-injecting function. These examples are remarkable evidence in support of the belief that the architectural plan for the complex macromolecules is present in the constituents themselves. There appears to be some selective procedure—perhaps through shape recognition properties—which raises the probability of assembly of the correct pattern of organic molecules to levels significantly greater than those arising from pure chance.[8]

Profound problems, however, remain. The living cell translates its genetic code by a molecular process of almost inconceivable complexity—a microscopic machine of at least fifty macromolecular components which are themselves coded in DNA. The code can only be translated by the products of the translation. It is a closed circle. When and how the circle became closed is an enigma. Attempts to prove that the structure of the code arose because of chemical affinities among certain amino acids have so far been negative. The alternative concept, that the choice of the code was arbitrary and has been enriched throughout history by a series of random choices, is not a scientific explanation.

Life exists on Earth; it emerged about 3 billion years ago. Before the event occurred, the chances that it would occur were exceedingly small. Furthermore, from our understanding of the probable manner in which our present atmosphere evolved, and of the critical part played by living organisms in this process, it seems that the decisive event—the transition from non-living to living matter—occurred only once, and could occur only once. The sequence of events from the solar nebula, through the barren Earth to the habitable atmosphere and the transition into life of inert matter, presents formidable odds against the emergence of human life. It is, indeed, a strange circumstance of contemporary science that the brilliant discoveries of recent years in the astronomical and biological sciences have revealed parallel universal mechanisms —one in the universal formation of planetary systems, and the other in the universal mechanisms that are basic to the properties of living organisms. It would be reasonable to anticipate that this progress would have helped to solve the problem of life's origins. On the contrary, the revelation of the immense complexity and delicacy of the physical and biological systems has transformed the question into realms of ever greater difficulty. In the confusion of our present understanding we may crave the philosophical certainty of Leibniz. Instead we may feel the uneasy neutrality of Descartes—"We know," he wrote, "that the world was created as in Genesis, but it is interesting to see how it might have grown naturally" —while as scientists we agree with Newton: "To tell us that every Species of Things is endow'd with an occult specifick Quality by which it acts and produces manifest Effects, is to tell us nothing."[9]

V

The Nature of the Universe

The application of modern science and technology to the observation of the Universe has revealed problems of an astonishing complexity. At present they are so far beyond the comprehension of the human intellect that elements of transcendentalism emerge as components of the explanation of the beginning of time and space.

Throughout history the transcendental element in cosmology has never been absent. Man has never obtained scientific answers to the problem of whether the Universe is finite or infinite, or how it began and how it will end. Nevertheless, the perspective of these questions and the nature of the possible scientific answers have constantly changed since ancient times. There have been three decisive changes in this perspective during this century, all resulting from events that occurred between 1918 and 1929. The three events were the consequence of observations of the heavens and there are few scientists today who do not accept their validity and their effect on our understanding of the nature of the Universe. They are, in their chronological order, the proof that the solar system is not the center of the Universe, evidence that the star system of the Milky Way is merely one unit among immense numbers of similar systems, and evidence for the expansion of the Universe.

THE SUN-CENTERED UNIVERSE

A strange feature of the history of cosmology is that the revolutionary Copernican doctrine did not substantially change man's wider concepts about the nature of the Universe. In Aristotle's cosmology the Earth was fixed at the center of the Universe and matter and space ended at the sphere of the stars. With the acceptance of the Copernican theory the Sun instead of the Earth became the center of the Universe. The center of the Sun coincided with the center of the finite stellar sphere.

Of course there were important changes in the philosophical outlook. In the Earth-centered universe it was the motion of the stellar sphere which provided the motive force for the movement of the planets. The stellar sphere defined an absolute frame of reference and objects moved towards the center of this sphere—the Earth—of their own accord. The Copernican universe deprived the stellar sphere of any such essential function. There could no longer be an absolute center toward which objects fell, since they fell toward an Earth that was in motion. There was no longer a necessity for the sphere of the stars to be in motion and the constraints that made it necessary for the stellar sphere to be finite were removed.

The concept of the infinite Universe which began to emerge was not new. Although Aristotle had given a logical proof that a void outside the stellar sphere was impossible and therefore the Universe was finite, his proof was not generally accepted. By the thirteenth century the popular view was in favor of an infinite universe beyond the sphere of the fixed stars. This infinite space contained no matter but was regarded as the abode of God and the angels. The integration of this concept of an infinite universe with the Copernican theory followed soon after the publication of *De revolutionibus*. In 1576 Thomas Digges published his *Perfit Description of the Caelestiall Orbes*. This work contains the first diagram of a universe in which the stars are no longer constrained to the surface of a fixed sphere but "This orbe of starres fixed infinitely and extendeth hitself in altitude asphericallye." The idea that stars could be scattered through infinite space received substantial support when Galileo discovered countless new stars with his telescope where none had been seen before. The ideas of inertia and gravity introduced by Descartes and Newton subsequently gave a scientific validity to the concept of an infinite universe. At the time of Newton's death in 1727 few would have queried the existence of a Sun-centered infinite universe populated by an infinite number of particles whose behavior was governed by the laws of inertia and gravity.

Today it seems remarkable that nearly two centuries elapsed before the idea of the Sun-centered universe was challenged. There were two problems which together greatly confused the issue. First there was the scientific problem: no one could measure the distance of the stars. This might not have been such a hindrance had it not been for the egocentric conviction that man must be near the center of the Universe. Indeed, a retrospective judgment is that the abandonment of the idea of the Sun-centered universe would not have been delayed until the twentieth century but for this egoism of man.

Soon after Galileo discovered that the Milky Way was composed of stars there was speculation about the apparent conflict between the idea of an infinite universe and the fact that this Milky Way stretched in a band around the heavens. An important step was taken by an English north-country teacher of navigation, Thomas Wright of Durham (1711–1786), who appears to have been stimulated to look at the heavens through a small telescope by a popular book *Astro-Theology* written by William Derham, who was a friend of Edmond Halley. It was clear to Wright that the stars in the Milky Way stretched right around the heavens and that there were far fewer stars in other directions. Since he felt it necessary to preserve the priority of the Sun, and the symmetry associated with a divine creator, he concluded that the stars were distributed throughout a giant disk extending infinitely outwards with the Sun at the center. Wright's hypothesis, published in 1750 in his book *An Original Theory or New Hypothesis of the Universe,* appears to be the first scientific argument that the stars in the heavens were not arranged with spherical symmetry around the Sun.

Wright died in 1786, when William Herschel was making his great series of observations of the heavens with the succession of telescopes which he contructed himself.[1] Since Herschel could not measure the distance of the stars, he made a catalogue of the brightness of a large number of them. Assuming that the apparent-brightness of one star differs from that of another because of their different distances from Earth, and not because of any inherent difference between them, he devised a scheme of stellar distribution, which bore some resemblance to the scheme of Thomas Wright. In Herschel's system the stars of the Milky Way were also arranged in the form of a disk. Unlike Wright, Herschel concluded that the stars did not extend to infinity, but rather that the disk in which they were contained had the shape of a rectangular box with a length five times its thickness.

By the end of the nineteenth century it was clearly evident that the stars were not distributed with spherical symmetry around the Sun but no one had the courage, or indeed the evidence, to suggest that the stars in the disk of the Milky Way might be asymmetrically distributed about the Sun. The decisive evidence could not be obtained until a means existed for measuring the distances of those stars, or groups of stars, which defined the shape and size of the Milky Way—and this did not occur until 1919.

In defending his conclusion that the Sun was situated at a great distance from the central region of the Milky Way, during the course of a famous meeting of the National Academy of Sciences in Washington on 26 April 1920, Harlow Shapley said:

One consequence of the cluster theory of the galactic system is that the Sun is found to be very distant from the center of the galaxy. It appears that we are near the center of a large local cluster or cloud of stars, but that cloud is at least sixty thousand light years from the galactic center. Twenty years ago Newcomb remarked that the Sun *appeared* to be in the galactic plane because the Milky Way is a great circle—an encircling band of light—and that the Sun also *appears* near the center of the Universe because the star density falls off with distance in all directions. But he concluded as follows: "Ptolemy showed by evidence, which, from his standpoint, looked as sound as that which we have cited, that the Earth was fixed in the center of the Universe. May we not be the victim of some fallacy, as he was?" The answer to Newcomb's question is: Yes, we have been victimized by the chance position of the Sun near the center of a subordinate system, and misled by the consequent phenomena, to think that we are God's own appointed, right in the thick of things.[2]

Shapley, who was then at Mt. Wilson, recognized the value of the globular clusters for any assessment of the structure of the Milky Way. These globular clusters are great conglomerations of stars approximately spherical in shape and containing 10^5 or 10^6 stars. They are all in one half of the sky, concentrated towards the band of the Milky Way, and a third of them are concentrated in a small area centered upon the great star cloud in Sagittarius. About one hundred are known and nearly all of these had been catalogued by Herschel before the end of the eighteenth century—it is therefore surprising that Shapley was the first astronomer to consider the implication of their uneven distribution. Shapley took the courageous step of assuming that the great clusters defined the general outline of the Milky Way, and with the means to determine the distances of the clusters from the solar system, he defined the nature of the local galaxy, or Milky Way, as we understand it today: a system of 100 billion stars distributed throughout a flattened disk 100,000 light years across, with the Sun in a spiral arm 33,000 light years from the central region.

There are remarkable features of this long trail to the dethronement of the Earth's position as the central feature of the Universe. The necessity of measuring the distances of the stars was long recognized to be an important matter for astronomers. The possibility of doing so by a trigonometrical method, using as a baseline the diameter of the Earth's orbit around the Sun, and observing the parallactic displacement of a star against the faint background of very distant stars, was also well understood. However, even the nearest stars were so distant that this parallactic displacement was too small to be measured by the instruments available in the eighteenth century. (The nearest star to the Sun has a parallax angle* of only 0.76 seconds of arc—corresponding to an image displacement of only 0.03 millimeters at the focus of a telescope having a focal length of 4 meters when the star is observed across the diameter of the Earth's orbit.) After many astronomers had attempted to measure stellar parallax, James Bradley, third Astronomer Royal of England, thought he had succeeded, in 1727, in measuring the distance of the star γ-Draconis, passing nearly through the zenith at Greenwich, by determining its exact position throughout the year using a telescope pointed at the zenith. He had found a large shift in position, of 40 seconds of arc, but it was not in the direction to be expected from the parallactic displacement of the star. He had discovered the phenomenon of aberration: light travels at a finite speed and when the light from a star reaches an observer on Earth, the relative positions of star and Earth are no longer the same as when the light was emitted. The aberration angle is therefore determined by the ratio of the Earth's orbital speed to the velocity of light.

Fifty years after Bradley's work, Herschel attempted to measure parallactic displacement using his superior telescopes. The existence of "double stars" had been known for

*The parallax is the displacement corresponding to the *radius* of the Earth's orbit around the Sun. A star at a distance of 1 parsec would subtend an angle of 1 sec of arc across the radius of the Earth's orbit. 1 parsec = 30.86 × 10¹² km or 3.26 light years.

the past century, and they were believed to be pairs of stars which fortuitously appeared close to one another in the sky. Herschel made precision measurements of the separation of these stars throughout the year in the expectation that the parallactic shift of one of the stars would be greater than that of its companion, believed to be more distant. Herschel certainly found displacement—but the changes were not those to be expected from parallactic displacement, and he concluded that he was observing pairs of stars whose coincidence in position arose because they were binary stars in which the two stars orbit around one another. For the second time, the attempt to measure the parallax had failed, but the investigation had led to unexpected discoveries.

In principle the determination of stellar parallax was simple, but it defeated all attempts at measurement for over two centuries after the invention of the telescope. Success was finally achieved by Friedrich Bessel in 1838. In 1810 the King of Prussia established a new observatory at Königsberg and summoned a brilliant young man—Bessel—to direct it. In twelve years Bessel determined the position of 75,000 stars with a precision hitherto unapproached. Then he asked J. von Fraunhofer in Munich to construct for him a special device known as a heliometer—in principle a large double-image micrometer. With this Bessel measured the position of the star 61 Cygni over a period of a year and found that its apparent position changed against the background of the faint stars. In 1838 he was able to announce the measurement of the first stellar parallax—only 0.30 seconds of arc representing a distance of eleven light years. It was a remarkable achievement, made possible by the combination of the superb instrument built by Fraunhofer and Bessel's own observational skills, of which he said, "the reduction by the observer himself of the results of his observations is an essential condition for success in all astronomical research."

This great distance of one of the nearer stars must have been a severe shock to those who clung to the belief that the Universe was not much larger than the solar system. The atmosphere of those days is vividly conveyed by the Presidential addresses to the Royal Astronomical Society on the subject of the award of the Society's Gold Medal in 1840. At the meeting of 14 February, the President, Sir John Herschel, gave a long explanation of the reasons which had led the Council of the Society to award the medal to Giovanni Plana, the director of the Turin Observatory, for his work on lunar theory, instead of to Bessel. After referring to the measurements of Bessel and Henderson* Sir John said that the "observations which [are,] it would appear, beyond question, have brought us to the very threshold of that long-sought portal which is to open to us a measurable pathway into regions where the wings of fancy have hitherto been overborne by the weight or baffled by the vagueness of the illimitable and the infinite." But after describing Bessel's measurements as they were available to the Society Sir John went on to say:

It may now be reasonably asked, If all this be so, why have your Council hesitated to mark this grand discovery with that distinct stamp of their conviction and applause, which the award of their annual medal would confer? A problem of this difficulty and importance solved, so long the cynosure of every astronomer's wishes—the ultimate test of every observer's accuracy—the great landmark and *ne plus ultra* of our progress, thus at once rooted up and cast aside, as it were, by a *tour de force*, ought surely to have commanded all suffrages. It is understood, however, that we have not yet all M. Bessel's observations before us. There is a second series, equally unequivocal (as we are given to understand) in the tenor, and leading to almost exactly the same numerical value of the parallax, and not yet communicated to the public. Under these circumstances, it became the duty of your Council to suspend their decision.

*Thomas Henderson, a Scotsman, had measured the parallactic displacement of α-Centauri during observations at the Cape of Good Hope Observatory in 1832–34. He became Astronomer Royal for Scotland in 1834 and did not announce his results until January 1839.

But, should the evidence finally placed before them at a future opportunity justify their coming to such a conclusion, it must not be doubted that they will seize with gladness the occasion to crown, with such laurels as they have it in their power to extend, the greatest triumph of modern practical astronomy.[3]

A year later Sir John and the Council were satisfied with Bessel's measurements and in his address on 12 February 1841 announcing the award of the gold medal to Bessel, Sir John paid tribute to the instrument "with which Bessel made these most remarkable observations . . . a heliometer of large dimensions, and with an exquisite object glass by Fraunhofer. I well remember to have seen the object-glass at Munich before it was cut, and to have been not a little amazed at the boldness of the maker who would devote a glass, which at that time would have been considered in England almost invaluable, to so hazardous an operation. Little did I then imagine the noble purpose it was destined to accomplish." In concluding his address Sir John said, "Gentlemen of the Astronomical Society, I congratulate you and myself that we have lived to see the great and hitherto impassable barrier to our excursions into the sidereal universe; that barrier against which we have chafed so long and so vainly—(*aestuantes angusto limite mundi*)—almost simultaneously overleaped at three* different points. It is the greatest and most glorious triumph which practical astronomy has ever witnessed."[4]

A hundred years before Bessel's measurement, Bradley had warned the astronomical community that his own measurements placed a limit of one second of arc on the parallax, but the implication that this placed the nearest stars many light years distant does not seem to have been absorbed into the stream of thought about the possible isolation of the solar system in space. Neither were astronomers prepared for the

---

*A reference to Henderson's measurements on α-Centauri and to those of W. Struve on α-Lyrae.

difficulty of extending the parallactic measurements to other stars. For the rest of the century many of the ablest astronomers wrestled with this problem. They were assisted by the introduction of photographic techniques into astronomy during the second half of the century, but in 1904 Simon Newcomb in his book *The Stars* was able to list only 72 stars for which parallax measurements had been made, of which 15 were listed as doubtful. It is remarkable that even in 1908, Jacobus Kapteyn, during an evening discourse at the Royal Institution, said that "with the exception of a hundred stars at most, we know nothing of the distances of the individual stars."[5] Already, at the time of Kapteyn's remarks, however, Frank Schlesinger was working on the problem, using the 40-inch refractor telescope at Yerkes with modern photographic techniques, and the number of known parallaxes soon increased. In 1924 he published a catalogue of 1,870 stars with measured parallaxes. Even with the best contemporary techniques the error of trigonometrical parallax measurements may not be less than about 0.01 seconds of arc, which means that such measurements at distances of more than one hundred parsecs are not reliable. The modern catalogues contain measurements on the distances of some ten thousand stars within this range.

The great difficulty and limitation of trigonometrical parallax measurements stimulated the search for other methods which could be applied to greater distances. One such method—that of secular parallax—makes use of the motion of the solar system through space. This motion shifts the observer on Earth by twenty kilometers per second and provides a continuously increasing baseline equivalent in one year to twice the diameter of the Earth's orbit around the Sun. The observed displacement of a star is compounded also with the proper motion through space of the star itself. Even so the method can be applied for a sufficiently large group of stars whose individual proper motions are such that the group as a whole may be taken to be at rest. Another method uses the spectrograph to assess the absolute magnitude* of the star from its spectral characteristics; then, by measuring the apparent magnitude, the distance can be estimated. These methods have been important in extending our knowledge of the distance of stars lying beyond one hundred parsecs from the solar system and have contributed directly to the method used by Shapley in his critical measurements on the globular clusters.

The major advance which was to be of such consequence to our understanding of the Milky Way and of the entire Universe followed from a discovery in 1908 by Henrietta Leavitt of Harvard, who studied the variation in intensity of certain types of stars in the Small Magellanic Cloud. These stars are known as Cepheid variables, a class named after the star δ-Cephei, which changes its brightness by a magnitude, from maximum to minimum, in five days. She noted the curious fact that the brighter stars took longer for their light to vary from maximum to minimum and, indeed, found that there was a definite relationship between the brightness of the star and its period of light variation. There were Cepheid variables among the stars whose distance could be estimated from the spectrographic-absolute magnitude method mentioned above, and Harlow Shapley, working with the 60-inch telescope at Mt. Wilson, calibrated Leavitt's apparent-brightness/period relationship in terms of absolute magnitude.** It was then possible, by measuring the period of a Cepheid variable, to find its absolute magnitude, and hence, by observing its apparent magnitude, to determine its distance by the inverse-square law.

Shapley proceeded to study the Cepheid variables in the globular clusters, and by 1918 had determined the distance of 25 of the 100 known objects of this type. He found them

---

*Throughout this book, "magnitude" is used in the astronomical context, as a measure of brightness.

**In 1955, Cepheid variables were discovered in the nearby "open" or "galactic" clusters. The Hydra cluster, for example, is close enough (about 42 parsecs) for the distance to be measured by the method of secular parallax.

to be at very great distances, mostly in the range of 15,000 to 100,000 light years from Earth and this, with his recognition of the significance of their uneven distribution over the sky, led to his bold description of the structure of the Milky Way and the removal of the Sun from a preeminent position in the Universe. Although there were continuing arguments about the magnitude of the distances assigned by Shapley, his general conclusions were quickly accepted. His results were published in 1918 and 1919 in the *Astrophysical Journal* and the Mt. Wilson Observatory *Contributions* and five years later it would have been difficult to find an astronomer who still defended the idea that the Sun might be close to the center of the Milky Way. The age-old egocentric concept of man's place in the Universe had been finally eradicated. It is curious to reflect that this general conclusion could have been substantiated qualitatively more than a century earlier. The evidence of Digges and Herschel on the distribution of the stars was sufficient to destroy the belief in the spherical symmetry of the star system, and Herschel had catalogued nearly all the globular clusters. No one, however, seems to have recognized the inescapable significance of their distribution over the sky until Shapley did so. Whether Shapley's arguments on this point would have been sufficient to overcome the belief of man in his centralized position if he had not been able at the same time to measure their distances is debatable.

THE PROBLEM OF THE NEBULAE

To the unaided eye large regions of the Milky Way appear as luminous clouds, but with even a small telescope many of these regions resolve into stars. Indeed, the success of Galileo led to the belief that all such luminous regions, or nebulae, would resolve into stars when viewed through a sufficiently powerful telescope. In the second half of the eighteenth century catalogues of those nebulae which were not resolvable into stars were published. In 1755 N. L. de La Caille published a catalogue, *Sur les étoiles du ciel austral,* of 42 such nebulae, and in 1784 C. Messier issued his famous catalogue, *Catalogue des nebuleuses,* of 103 objects which are identified by his name today. One year later Herschel published the first of the catalogues resulting from observations with his great series of telescopes—and listed a thousand nebulae. In that year, 1785, Herschel read his paper, "On the Construction of the Heavens," to the Royal Society. He said that as the power of the telescope was increased an observer "perceives that those objects which had been called nebulae are evidently nothing but clusters of stars. He finds their number increase upon him, and when he resolves one nebula into stars he discovers ten new ones which he cannot resolve." Eventually Herschel catalogued 2,500 nebulae. He believed that all the nebulae were composed of stars and that they were at great distances outside the confines of the Milky Way system.

Herschel's conclusions about the nature of the nebulae could not be easily substantiated because the means of measuring the distances did not exist. Furthermore, although his telescopes had revealed so many hitherto unknown objects of this type, many of which he had resolved into stars, he was not the first to suggest that star systems existed in space outside the confines of the Milky Way. The first observer of the heavens to make this suggestion appears to have been Thomas Wright in 1750, in the book to which reference has already been made in connection with his ideas on the distribution of the stars in the Milky Way. Wright's book does not appear to have been well known—Herschel possessed a copy but makes no reference to Wright's ideas—but a summary of his hypothesis in a German periodical stimulated Immanuel Kant to write his *General Natural History and Theory of the Heavens.* This work,* published in 1755 and based on the principles of Newtonian theory, proposed the nebular hypothesis for the formation of the solar system. On the problem of the nebulae, Kant enunciated his theory of the island universes:

---

*Allgemeine Natur—Geschichte und Theorie des Himmels, originally published anonymously in 1755.

Their analogy with our own system of stars; their form, which is precisely what it should be according to our theory; the faintness of their light, which denotes an infinite distance; all are in admirable accord and lead us to consider these elliptical spots as systems of the same order as ours —in a word, to be Milky Ways similar to the one whose constitution we have explained . . . a vast field lies open to discoveries, and observation alone will give the key. . . .

Although Kant gave a concise enunciation of the "island Universe" theory of the nebulae, the concept, as a philosophical idea, had in fact been proposed twenty years earlier by the Swedish mystic Emmanuel Swedenborg, who in 1734 suggested that there existed in space many stellar systems of about the same size and extent as the Milky Way.[6] Similar intuitive ideas were published by J. H. Lambert in 1761. In his *Cosmological Letters (Cosmologische Briefe über die Einrichtung des Weltsystems)* Lambert considered the unity and constitution of the Universe. His hypothesis of an obscure central body around which the Sun and stars gravitated to form the clusters making up the Milky Way was extended to the idea that a still larger system existed made up of many Milky Ways. Nevertheless Kant was indubitably correct when he asserted that "observation alone will give the key"; but neither Kant nor the observers of the eighteenth and nineteenth centuries realized the complexity of the problem of the nebulae.

Always in those centuries the issues were confused by the astronomers' inability to measure distances. Bearing in mind that success was not achieved until the 1920s it is remarkable that in 1767 John Michell applied statistical methods to this problem by comparing the dimensions of the nebulae with our own system in order to deduce the magnitude of the brightest stars in the nebulae[7] He had to base his arguments on the assumption that, on the average, the stars were equal in absolute magnitude and size to the Sun. Even so, his conclusion that the brightest stars of a nearby galaxy would have a visual magnitude of 13.8 was not far wrong, since we know today that the brightest stars in the nearest galaxies are only about 2 magnitudes fainter. However, the arguments and uncertainties continued as to the true nature of the nebulae. In the years when Herschel first turned his attention to the problem, and as he achieved success in resolving many of the nebulae into stars when he increased the power of his telescopes, he became convinced that the nebulae were star systems—island universes—distributed in space outside of the Milky Way. Although with his 20-foot telescope there were many nebulae which he could not resolve into stars, he was certain that he would be able to do so with the greater resolving power of the 40-foot telescope which he was planning—and that these unresolved nebulae were large enough to "outvie our Milky Way in grandeur."[8] Herschel never succeeded in this ambition, and the arguments about the nature of the nebulae were immensely involved because of the disagreements over the extent of the Milky Way system. Nevertheless, there seems little question that at the age of 76 Herschel changed his mind to a certain extent toward the conclusion, which we recognize today as the correct one, that the term nebula had been assigned to two quite different kinds of objects. Some are extragalactic, resolvable into stars; others are luminous gaseous nebulosities within the Milky Way system. In 1814, Herschel made the comment that "other objects there are, where a great space-gathering power will only increase the brightness of the nebulosity." He had discovered a number of stars that seemed to lie in the midst of some of these nebulosities, "a shining fluid, of nature totally unknown to us."[9]

It was over a century before the confusion could be dispelled by the actual measurement of the distance of those nebulae which were resolvable into stars. In the meantime the contrary views of the powerful intellects who followed Herschel provide a classic example of man's attempt to clarify an extremely difficult problem additionally confused by nature—and incidentally underline the brilliant achievement and deep insight of Herschel.

The most outstanding practical effort to pursue Herschel's concepts was made by the Third Earl of Rosse at Birr in Ireland. The mirror of Herschel's largest telescope had an aperture of 48 inches. Lord Rosse succeeded in manufacturing a reflector with a diameter of 72 inches, and for three quarters of a century this remained the largest telescope in the world. In 1845 he used it to study the nebulae. Like Herschel, he failed to resolve the Orion nebula into stars but made the famous discovery that many of the other nebulae which were resolvable into stars had a marked spiral structure. Lord Rosse's drawing of the face-on spiral nebula in Canes Venatici placed alongside a contemporary photograph taken with the 200-inch telescope on Palomar illustrates his superb observational powers. Yet Rosse concluded that this system, an extragalactic system over 10 million light years distant, and others like it were simply clusters of stars within the Milky Way!

Indeed the second half of the nineteenth century is marked by a reaction against the opinions of Kant and Herschel. In 1858 Herbert Spencer illustrated the confusion of thought existing at that time when he wrote in support of the idea that all the nebulae lay within the Milky Way. "When to the fact that the general mass of the nebulae are antithetical in position to the general mass of the stars, we add the fact that local regions of nebulae are regions where stars are scarce, and the further fact that single nebulae are habitually found in comparatively starless spots, does not the proof of a physical connection become overwhelming?"[10] Of course Spencer knew nothing of the light absorption by the dust clouds in the Milky Way which is the cause of the "antithetical" position of the nebulae with respect to the stars, nor that there were two quite different kinds of nebulae.

The idea of extragalactic systems, or island universes, suffered a far more serious setback when William Huggins, a London mercer with no formal scientific training, decided to use the new technique of spectroscopic analysis to investigate the nebulae. After studying the spectral lines from some of the brighter stars, in 1863 Huggins turned his instrument to the nebulae. In a famous evening discourse to the Royal Institution of London on 19 May 1865 he said,

> Besides the stars, the heavens are mottled over with feebly shining cloud-like patches and spots, often presenting strange and fantastic forms. Between five thousand and six thousand of these so called *Nebulae* are known. What is the nature of these strange objects? Dense swarms of Suns melted into one mass by their enormous distance? Chaotic masses of the primordial material of the Universe? The telescope alone would fail to give the answers to these questions and the analysis of objects so feebly luminous appeared hopeless. In August last the speaker directed his telescope, armed with the spectrum apparatus, to a small but comparatively bright nebula, 37 H iv. His surprise was great to observe, that in place of a band of coloured light, such as the spectrum of a star would appear, the light of this object remained concentrated in three bright bluish-green lines, separated by dark intervals. This order of spectrum showed the source of the light was luminous *gas*. [11]

The conclusions of Rosse, who had observed the nebulae with the largest telescope in the world, and this work of Huggins, who had fortuitously chosen a nebula which actually was luminous gas within the confines of the Milky Way, seemed to settle the argument against the extragalactic concept. A distinguished historian, Agnes Clerke, wrote at the end of the century that the question whether nebulae were external galaxies "hardly any longer needs discussion. . . . No competent thinker, with the whole of the evidence before him, can now, it is safe to say, maintain any single nebula to be a star system of co-ordinate rank with the Milky Way." [12]

The arguments about the position of the Sun in the Milky Way reached a climax in the "great debate" between Harlow Shapley and Heber Curtis of the Lick Observatory at the National Academy of Sciences in Washington on 26 April 1920, to which reference has been made. In 1917 Curtis had compared the maximum magnitudes of five novae in four spiral nebulae with those of novae in the Milky Way. He concluded that the spiral nebulae were about 20 million light years beyond the Milky Way. It is curious that Shapley, who had been correct about the Sun's position, opposed this view and maintained that the nebulae were either within the Milky Way system or close neighbors. Although Curtis had made mistakes in his distance estimates (some of the novae were later found to be supernovae) he was correct in his opinion that the spirals were distant star systems. Further, the discovery of the luminosity-period relationship of the Cepheid variables, which had enabled Shapley to settle the dispute about the structure of the Milky Way, enabled Edwin Hubble a few years later to reach unambiguous conclusions about the distances of the spiral nebulae.

In 1926 Hubble published the results of his observations on the nebula catalogued as M33. He had been able to resolve this nebula into stars, among which he identified Cepheid variables, and thus, as Shapley had done for the globular clusters, he was able to measure the distance to M33 by using the period-luminosity relationship. He thereby produced the first unambiguous proof that a nebula was an extragalactic star system remote from the Milky Way. In his paper on M33 he referred to similar measurements on M31, the great spiral nebula in Andromeda (a naked-eye object under favorable conditions), a nebula 2 million light years distant, which we now believe to be almost identical to the Milky Way in size, structure and stellar content.

Hubble's distance measurements were made possible because the new 100-inch telescope on Mt. Wilson, with a light-gathering power superior to that of any previous telescope, enabled him to identify the Cepheid variables in the stars comprising these nebulae. However, even with this telescope there were only a very few nebulae in which the Cepheids could be measured. He therefore made a statistical study of four hundred nebulae, making the assumption that nebulae of the same type would have the same absolute magnitude. On this basis he estimated the distances and numbers of the nebulae whose type he could identify, but which were too distant for the measurement of Cepheid variables. At that time, in 1926, he estimated that the 100-inch telescope, when photographing to a limiting magnitude of +18, would encompass 2 million extragalactic nebulae, to a distance of 140 million light years.

When Hubble established the extragalactic nature of the nebulae, he found that they were essentially of two types. The spherical and elliptical galaxies, exhibiting little or no structure, represented about 20 percent of the galaxies he observed. Apart from a few percent with irregular structure, the remainder were of spiral formation. The belief that these types of nebulae formed the Universe, however, did not long survive the application of the new techniques of radio astronomy after World War II. In 1951 some of the newly discovered radio sources were identified with hitherto unrecognized objects in the Universe, radio galaxies, and a decade later the quasars were discovered. These objects are compact and emit large amounts of energy. Their true nature, their place in the evolutionary sequence and the processes by which such great energies are emitted from regions of space which are small by the standard of normal galaxies, are not yet understood. For our present purpose it is important to remark that the cooperation of the radio and optical astronomers in this work has extended our penetration into the Universe to immensely greater distances than envisaged by Hubble in 1926. The introduction into astronomical research of the 200-inch Hale telescope on Palomar in 1949 extended the penetration to about 2 billion light years, and today the most remote quasars have been identified at distances of probably more than 6 billion light years.

How many extragalactic nebulae of all these various types lie within the field of view of modern telescopes? Some idea of the immense number can be deduced from the sky photographs taken by the modern 48-inch Schmidt telescope at Siding Spring in Australia. The faintest object photographed by this instrument has a magnitude of $+23$ and about twenty-five hundred extragalactic objects can be counted in a square degree of the sky.[13] There are 41,250 square degrees in the whole sky; thus to the twenty-third magnitude there are about 100 million galaxies. At this sensitivity limit the number of observable objects is still increasing by two or three times per magnitude. With larger telescopes and modern instrumentation the sensitivity limit can probably be improved by about 2 magnitudes—an improvement apparently already achieved by the new Soviet 236-inch telescope in the North Caucasus.[14] In this case some 500 to 1,000 million extragalactic objects are within the observational limits of modern telescopes on Earth. If the plans proceed for placing a large space telescope in orbit in 1983, using the US Shuttle, then the remaining constraints imposed by the Earth's atmosphere will disappear, and a further extension of the sensitivity limit by several magnitudes can be envisaged. Will it be found that the numbers of extragalactic objects continue to increase at the rate of two or three times per fainter magnitude? For how long this increase continues is one of the fundamental unanswered problems in cosmology.

In these studies of the extragalactic Universe the problem of distance measurement remains a severe difficulty—as it has done throughout the centuries in the development of our understanding of the Universe. The use of Cepheid variables as distance indicators is very limited, of the order of a few million light years. Within that range, and especially within the local group of galaxies, the distances obtained from the Cepheid measurements may be compared with results obtained by a number of other methods, for example with the distances derived from measurements of the magnitude of the globular clusters, novae and supernovae. The general agreement of these values engenders confidence that, at least to distances of the order of a few million light years, the distance scale may be considered to be reasonably established. Some authorities believe that in cases where Cepheids can be identified, the errors are no greater than $\pm$ 15 percent. Unfortunately only a very few extragalactic systems lie within this distance scale, while beyond, the estimates of distance become increasingly uncertain. All such determinations of extragalactic distances are ultimately based on the assumption that recognizable types of distant objects are similar to nearby objects of the same type. Estimates of the apparent magnitude of supernovae occurring in distant galaxies, and of globular clusters, have been used to extend the distance scale to about 150 million light years. Beyond that order of distance the estimates depend on the type of assumption made by Hubble in his 1926 statistical analysis of four hundred galaxies: that galaxies of similar type will have similar characteristics, such as luminosities and diameters, and that by measuring the apparent magnitudes or diameters, the distance can be derived. By this method, either for individual galaxies recognizable in the clusters or, at greater distances, for the whole cluster, the distance scale has been extended to more than a billion light years. Beyond that the phenomenon of the red-shift, now to be considered, is the criterion on which the scale of the Universe has to be judged.

THE EXPANSION OF THE UNIVERSE

Hubble published the results which settled finally the existence of galaxies external to the Milky Way in 1926. Three years later he published his evidence that the Universe was not static but was expanding.

Whereas the existence of extragalactic systems had been debated for centuries, no one seems to have considered the possibility that the Universe might exhibit phenomena of large scale movement. Velocities, small compared with the speed of light, of bodies in the solar system and stars in the Milky Way were well known, but the interpretation of Hubble's new results on the galaxies implied a cosmical expansion of the Universe with very high velocities. Further, the evidence showed that the velocity of expansion increased linearly with the distance.

The curious theoretical and observational history of this surprising phenomenon was contained within two decades. In 1917 Einstein published his general theory of relativity in which he sought a generalization of the principle that the inertial properties of matter are determined by the behavior of matter on a cosmic scale. It was understandable that he should seek solutions of the equations which would yield a static universe, and it appeared that he had done so successfully by the introduction of a new constant into the equations. In this solution of Einstein's the radius of the Universe was simply related to the mass and to the density—and it was static. However, within a few months Willem de Sitter[15] showed that Einstein's solution was not unique, and that on the basis of the equations of general relativity and making the same assumptions, there was another solution. It was a solution which predicted a remarkable model for the Universe: a universe which was static only as long as it was empty of matter. If a particle, or nebula, were introduced into the model then it would not remain at rest, as in Einstein's solution, but would acquire an ever-increasing velocity away from the observer. Five years later the Russian mathematician A. Friedmann showed that there was a whole family of cosmological models for the Universe, based on general relativity, in which the mean density of matter varied with time.

These predictions of general relativity introduced once more into human thought the conflict with common sense so clearly in evidence at the time of the Copernican hypothesis in the sixteenth century. Again there were compelling reasons for accepting the new concepts. At least the Einstein universe was finite and provided a scientific and philosophical escape from the infinities associated with the Newtonian universe. On 19 May 1919 the total eclipse of the Sun provided the means of verifying Einstein's prediction of the amount of deflection of starlight grazing the Sun's disk. In 1925 W. S. Adams, using the 100-inch telescope on Mt. Wilson, verified Eddington's prediction that, on the basis of Einstein's theory, the spectral lines of the white dwarf star Sirius B should be displaced towards the red (increased in wavelength) because of its immense gravitational field, which is thirty thousand times that of the Earth. To this extent, that the Einstein theory had been found to be successful in these predictions, as well as in providing an explanation of the shift in the perihelion of the planet Mercury, the observational evidence for the predicted expansion of the Universe may have been more readily apprehended.

The observational foundations for the conclusions reached by Hubble in 1929 had been laid much earlier by V. M. Slipher at the Lowell Observatory, Flagstaff, Arizona. In 1912 he obtained a spectrum of the M31 spiral nebula and found that the spectral lines, which could be identified from certain elements, were not in their correct position on the spectrograph. The shift was towards the violet end of the spectrum—that is, their wavelength was shorter than the wavelength of the identical lines from the same elements in the observatory. The obvious cause of such a shift in wavelength is that the source is in motion with respect to the observer—the Doppler effect —and a shift towards the violet, or shorter wavelength, indicates that the source and observer are approaching each other. Slipher found that the shift implied a velocity of approach of 190 miles per second.

Because of the low surface brightness of the nebular images, these measurements were very difficult, and in another two years of observing, Slipher had succeeded in measuring only another thirteen velocities. It soon appeared, however, that the result on M31 was unusual in that the majority of measurements showed that the spectral lines were displaced towards the red—implying increasing separation of source and observer. By 1925 forty-one nebulae had been measured by Slipher and measurements on another four were available from elsewhere. In 1929 Hubble had forty-six spectral measurements available. Among these he had distance measurements for eighteen isolated nebulae and for the Virgo cluster. It was on the basis of this information that he found the remarkable relationship between the displacement of the spectral lines of a galaxy to the red and the distance of the galaxy.[16] This implied that the apparent velocity of recession of a galaxy was simply and linearly related to its distance from the observer. Since Slipher's measurements, interpreted as a Doppler effect, implied that in this sample velocities of recession up to 1,200 miles per second were involved, it is understandable that Hubble exhibited caution in drawing the conclusion that real velocities of recession were involved.

In 1928 M. L. Humason began his measurements of the spectra of the nebulae using the 100-inch Mt. Wilson telescope, and in the 1931 publication of Hubble and Humason, the linear law was shown to extend to distances of about 100 million light years where the implied recessional velocities were 2,000 miles per second. The authors conclude: "The present contribution concerns a correlation of empirical data of observation. The writers are constrained to describe the 'apparent-velocity-displacements' without venturing on the interpretation and its cosmologic significance."[17]

Humason continued his measurements, and by 1935 had measured 150 new spectra of nebulae out to distances thirty-five times greater than the Virgo cluster—estimated to be 230 to 240 million light years. The redshift of the spectral lines of these distant nebulae implied recessional velocities of 26,000 miles per second or one seventh of the velocity of light. The linear law between the distance of a nebula and its apparent velocity of recession still held. In his Silliman memorial lectures at Yale in 1935,[18] Hubble said this:

> The necessary investigations are beset with difficulties and uncertainties, and conclusions from data now available are rather dubious. They are mentioned here in order to emphasize the fact that the interpretation of redshifts is at least partially within the range of empirical investigation. For this reason the attitude of the observer is somewhat different from that of the theoretical investigator. Because the telescopic resources are not yet exhausted, judgment may be suspended until it is known from observations whether

or not redshifts do actually represent motion. Meanwhile, redshifts may be expressed on a scale of velocities as a matter of convenience. They behave as velocity-shifts behave and they are very simply represented on the same familiar scale, regardless of the ultimate interpretation. The term 'apparent velocity' may be used in carefully considered statements, and the adjective always implied where it is omitted in general usage.

More than forty years have passed since Hubble's Silliman lectures, during which very great extensions have been made to the measurements of the redshift from extragalactic nebulae. The commissioning of the 200-inch Hale telescope in 1949, with its much greater light-gathering power, led to redshift measurements of extragalactic nebulae estimated to be at distances of a few billion light years.* For example, a cluster of galaxies in the constellation of Hydra, at a distance estimated to be 2 billion light years, showed a displacement of the spectral lines towards the red which, when interpreted as a velocity effect, implied a recessional velocity of 38,000 miles per second.

The discovery of the radio galaxies gave rise to even deeper penetrations into space, and in 1960 R. Minkowski used the 200-inch telescope to photograph a cluster of galaxies in the vicinity of one of these strong radio-emitting objects in the constellation of Boötes. In an area three minutes of arc in diameter he found sixty galaxies to the limiting magnitude of 21. One of the brightest of these galaxies was precisely in the position of the radio source. From the apparent magnitude of the galaxies the distance was estimated to be 4.5 billion light years. A spectral line identified as an oxygen line of wave-

---

*At about the time when the 200-inch telescope was coming into use, W. Baade working at Mt. Wilson discovered that the Cepheid variables in the extragalactic nebulae belonged to a different class of star (Population I) that was four times brighter than the Cepheids (belonging to Population II) which had been used in the Leavitt-Shapley calibrations. This had the effect of increasing all previously derived extragalactic distances by a factor of two. Baade announced this result, which doubled the distance and time scales of the Universe, at the 1952 meeting of the International Astronomical Union in Rome (Trans. IAU 8, 397, 1952).

length 3,727 angstroms was shifted to 5,447 angstroms, implying a recessional velocity of 86,000 miles per second or 46 percent of the velocity of light. The linear relationship between redshift and distance, estimated on the basis of apparent magnitudes of galaxies, was found to be maintained throughout this vast scale of distance.

The caution expressed by Hubble in his Silliman lectures has, over the years, been superseded by the almost unanimous opinion of astronomers that these redshifts are a cosmological effect associated with the expansion of the Universe. It cannot be said that a similar unanimity exists over the meaning of the far greater redshifts found in the objects known as quasars, the blue, radio-emitting objects, originally believed to be stars in the Milky Way but which were discovered in 1963 to have very large redshifts. Since that time several hundred quasars have been identified and their redshifts measured. Most of them have redshifts greater than that of the radio galaxy in Boötes, and at the upper limit a few are known with redshifts which imply a velocity of recession of the order 80 to 90 percent of the velocity of light.

Out to the most distant radio galaxy the extension of the distance scale has been based successively on the recognition of distant objects, similar to nearby objects, whose distance had been established by other means. No such possibility exists in the case of the quasars since, at least to the present day, they appear as a new class of object not conspicuously related to other components of the Universe whose distances are known by other methods. The distance scale for quasars has therefore been wholly based on the assumption that the redshift is a cosmological effect associated with the expansion of the Universe. Whether the Hubble linear relationship is maintained for such great redshifts is a critical matter of contemporary concern—small departures from linearity at these redshifts lead to markedly different conclusions about the past and the future history of the Universe. The quasars themselves do not contribute significantly in this assessment because there is a large scatter in the measurements. That is, quasars with the same apparent magnitude have widely differing redshifts. Nevertheless, within the limits that are considered to be feasible departures from the linear law, it is possible to estimate the distances of the quasars on the assumption that the redshifts do indeed arise wholly from the cosmological expansion of the Universe.

The answer, that nearly all quasars lie at distances greater than about 5 billion light-years, needs further qualification, however, since there is current dispute about the value of the Hubble constant—that is, the precise relation between the redshift and distance. The value obtained by Hubble was 530 kms per sec per megaparsec.* The revision of the distance scale and the extension of the measurements to greater redshifts showed that this value was about five times too large, and a figure of 100 kms per sec per megaparsec was considered to be appropriate. This implies that the "beginning" of the expansion was 10 billion years ago. Today, however, some authorities argue that the value is half this. In 1971 Allan Sandage in a lecture at the Royal Greenwich Observatory presented his latest results and gave the value of the constant as 55 kms per sec per megaparsec, implying a beginning of the expansion about 17 billion years ago. He claimed an accuracy of 15 percent but said, "It is sobering and is the better part of prudence to recall that but twenty years ago the value of the Hubble constant was thought to be 530 kms per sec per megaparsec, also with a stated accuracy of 15 percent."[19]

Rather than attempt to place a meaning on a distance for the quasars with the greatest known redshifts which, depending on the value taken for the Hubble constant, may be calculated numerically as about 8 billion or 15 billion light years, it is perhaps more important to inquire how far we look back into the past history of the Universe when we study these objects. If the greatest known quasar redshifts are a wholly

---

*That is, per million parsecs or 3.258 million light-years. Although these rates are very high on the cosmical scale, they are minute by local standards. This value of 530 kms per sec per megaparsec is equivalent to only 50 cms per sec at the nearest star.

cosmological expansion effect (and it must be said that the current astronomical literature is sprinkled with arguments against this assumption), then they existed one or two billion years after the beginning of the universal expansion. Few would doubt that in studying the distant clusters of galaxies such as the radio galaxy in Boötes we look back nearly halfway to the beginning of the expansion of the Universe, and on present evidence the majority of astronomers would accept that in viewing the most distant quasars the look-back time is more than 80 percent of the time to the beginning.

VI

The Origin of the Universe

Present-day discussion about the value of the Hubble constant—whether it is as low as 50 or as high as 100 kms per sec per megaparsec—is an argument about the time scale and the distance scale of the Universe, an argument about whether the beginning of the expansion was 20 billion or 10 billion years ago. These uncertainties of a factor of two do not obscure the fundamental evidence that the Universe is expanding. There would seem to be an inescapable logical conclusion from this evidence that the beginning of the expansion was of a universe totally different from that which we view today—that it must have been localized, and of immense density. The intellectual difficulty presented by this conclusion is severe and raises the question of whether the beginning of the expansion was also the beginning of the Universe—and in that case, what do we mean by the beginning? Because of these difficulties a section of the astronomical community has argued that the expansion of the Universe does not imply that there was a beginning of the expansion. However, in recent years fresh evidence has accumulated that seems to place new constraints on the discussions. For convenience in this chapter a Hubble constant of 100 kms per sec per megaparsec will be assumed, implying a beginning of the expansion 10 billion years ago, although this assumption in no way affects the basic considerations.

THE MICROWAVE ISOTROPIC BACKGROUND RADIATION

If the Universe is expanding, then it would seem natural to assume that the expansion is from a different earlier state. Since inevitably when we study the distant galaxies we can do so only by receiving the light, radio waves or other radiations which they emit, and since these radiations travel with the finite speed of light (186,000 miles per second), we have the ability to view the Universe as it was in past epochs. We see a galaxy at a distance of a billion light years as it was a billion years ago. This "look-back time" to the beginning of the expansion increases with the redshift. The precise relationship depends on the departure of the Hubble line from linearity, and this remains a critical matter to be determined. However, the limits of departure are reasonably well established. Reference has been made to the radio galaxy in Boötes as probably the most distant non-quasar object so far studied. The redshift is 0.461, implying a recessional velocity of 46 percent of the velocity of light. The light from this galaxy has taken 0.473 of the time since the beginning of the expansion and the possible departures from this value are not great. In other words, when observing this object we look back nearly halfway to the beginning of the expansion. Most professional astronomers would agree with this conclusion. And, as we have seen, if the redshifts of the quasars are due to the cosmological expansion of the Universe, then for those with the greatest redshifts so far measured the look-back time is 80 to 90 percent of the time to the beginning of the expansion.

The consequences seem clear. If the Universe is evolving in this way, then these observations of the distant galaxies should provide evidence that the condition of the Universe was indeed different in those earlier epochs. Many attempts have been made to find convincing evidence for these evolutionary differences, which might be expected to be manifest in a number of ways. Perhaps the most obvious is that we should expect to find the spatial density of the galaxies increasing as we view earlier epochs—that is, the number of galaxies per unit volume of space would be greater in the past than they are in the present epoch. When radio galaxies were discovered, it was realized that it was relatively easy to detect radio emissions from distant galaxies which were faint and difficult to photograph with the large optical telescopes. It was therefore argued that the counts of the number of radio sources (although unidentified) plotted as a function of their intensity (equivalent to distance) should be a test of the evolutionary theory. And, indeed, evidence was presented that the spatial density of these sources increased with decreasing intensity.

This apparently powerful evidence in favor of an evolutionary universe which has expanded from a beginning of high density was hotly disputed on various grounds, especially by an eminent group of astronomers who had evolved a different cosmological theory, known as the steady state or continuous creation theory. This theory, particularly associated with F. Hoyle, H. Bondi and T. Gold, emerged in 1948, stimulated by two separate considerations. First, the evolutionary theory implied an initial condition of high density and the concept of a beginning, which seemed to lie outside the scope of physical inquiry. Second, the value of the Hubble constant then extant (530 kms per sec per megaparsec) gave a time scale of only about 5 billion years from the beginning of the expansion, which was in conflict with the established age of the Earth, about 4.5 billion years. By no conceivable processes of condensation from the primeval gas could the solar system be considered to have emerged in the very early history of the Universe.

The revision of the distance scale by Baade in 1952 doubled the time scale for the evolution of the Universe and removed the second difficulty, but the steady state theory remained an attractive proposition because of the first consideration. The theory of the steady state involves the cosmological principle that on the large scale the Universe is unchanging and uniform. Since the Universe is expanding, the principle implies that new matter must be in process of continuous creation in order to maintain a constant density. As the nebulae move apart, so new ones are formed from the created matter at the appropriate rate to present an unchanging aspect to an observer anywhere in time and space. It is clear that in the observational test mentioned, that is, the spatial density of the galaxies as a function of epoch, this theory makes a contrary prediction to the evolutionary theory. In the evolutionary theory, the spatial density must increase with past epochs; the theory of the steady state predicts a constant spatial density at any epoch.

Although the number counts of the radio galaxies seemed to be in favor of the evolutionary theory, there were a number of difficulties inhibiting complete acceptance of these and other observations as conclusive evidence. This situation has changed significantly in recent years as a consequence of observations made during 1964 and 1965 at the Bell Telephone Laboratories in Holmdel, New Jersey. The discovery was made accidentally with radio receiving equipment of high sensitivity, which was working in the centimeter waveband for space communication tests using the Echo balloon satellite. In order to avoid effects of radio and other emissions from the ground a special horn-type aerial system had been developed so that signals received outside the main beam, in side-lobes, were much weaker than those received with the more conventional paraboloid. In spite of these precautions it was found that the noise level in the equipment when directed at the sky was one hundred times greater than it should have been, judging from the known sources of galactic or extragalactic radio emission at those wavelengths. Further, A. A. Penzias and R. W. Wilson, who were primarily concerned with these measurements, discovered that this excess noise was isotropic over the sky to a few percent. They announced that at a wavelength of 7 centimeters there existed an isotropic background radiation equivalent to a black body temperature of 3.5°K (later modified to 3.1°K and finally to 2.7°K). Several other workers soon confirmed this result at neighboring wavelengths and the effect became known as the "3 degree background radiation."

A remarkable feature of this discovery was that Professor R. H. Dicke and his colleagues at Princeton had been simultaneously considering the implications of a high temperature collapsed phase of the Universe, either as a singular event or as the epoch of maximum collapse in an oscillating universe. From considerations of the present density of the Universe they had concluded that the temperature at such an epoch would be 10^{10} deg K, and that it should be possible now to detect relic black body radiation which they calculated would have a temperature less than $40°K$. They had built apparatus working on a wavelength of 3 centimeters to attempt to find such radiation, but before results were obtained they learned of the successful measurements of Penzias and Wilson at Holmdel.[1]

In 1946 G. Gamov had proposed that the Universe "began" with a "hot big bang" and that relic radiation should now be observable at a temperature of $25°K$. Penzias and Wilson believed that they had detected this relic radiation, although the effective temperature they measured was ten times less than Gamov's prediction; his calculations were, in any case, sensitive to assumptions about the early helium production. Naturally, the discovery of the isotropic background radiation and its interpretation in terms of relic radiation led to a great debate, and some astronomers sought alternative explanations. However, measurements made from rockets in 1973 and balloons in 1974 established that the intensity of this isotropic background fitted the curve of black body radiation down to wavelengths of less than 1.5 millimeters (that is, below the peak intensity of the black body radiation appropriate to a temperature of $2.7°K$). Today this microwave isotropic background radiation is regarded as decisive evidence in favor of the evolutionary universe, and furthermore, as indicative of a high temperature state of the order 10^{10} deg K in the initial moments of the expansion.

THE PHYSICAL CONDITION OF THE INITIAL STATE

The theory of general relativity was accepted half a century ago and shortly thereafter Hubble found the observational evidence in favor of the non-static expanding universe. In recent years one or two variants of the general theory have been proposed, but whenever the predictions have been tested observationally the results have supported the Einstein theory. The most recent example occurred in 1976 when precise timing measurements were made on the signals from the Viking spacecraft as the planet Mars was in near conjunction with the Sun. These measurements gave very accurate values for the amount of deflection of radio waves grazing the solar disk and confirmed the Einstein value rather than a slightly different figure predicted by the competing theories. In the case of Hubble's evidence, no one has yet been able to suggest an acceptable alternative to the interpretation of the redshift in terms of a cosmological expansion of the Universe. The lingering doubts concerning the nature of the quasar redshifts do not invalidate this statement since, at least to a value of half the velocity of light, the interpretation is not disputed. Hence we find the overwhelming weight of theoretical and observational evidence in favor of a universe which is expanding and is compatible with the theory of general relativity. The theory of the steady state was an effort to embrace these two constraints but at the same time to avoid the consequence that the Universe was in an evolutionary state from a significantly different physical condition in past epochs. Since the proof in 1973–74 that the microwave radiation really does have the key characteristics of black body radiation, it appears that the theory of the steady state is not viable.

With the demise of the steady state concept, theorists have concentrated great attention during the last few years on the problems of the highly condensed state, which seems now to be an inescapable conclusion about the early condition of the Universe. Because of undetermined constants in the world models of general relativity, a variety of theoretical models for the evolutionary Universe can be derived. In some the Universe is predicted to expand for ever; in others it is cyclical, the expansion being followed by a contraction to another state of high density. All have in common the problem of the initial state of high density, for all predict an expansion from zero radius at time zero. As formulated at present the theory of the expanding Universe has this mathematical singularity as the starting point. Physically this would imply that at the beginning of the expansion the Universe must have been infinitely small and infinitely dense.

The occurrence of these infinities in the solution of the equations has been recognized since the original work of A. Friedmann and others on the cosmological solutions of the equations of general relativity more than fifty years ago. There has, however, been a tendency to disregard the problem on the grounds that the singularity arises because of assumptions that the Universe was exactly isotropic in the early state and that the difficulty might be a mathematical one, not having real relevance to the physical condition of the Universe near time zero. In addition to providing observational evidence that the primeval material of the Universe was concentrated in a condition of high density and temperature, the discovery and measurement of the microwave background radiation have had an important theoretical consequence with respect to the problem of the initial singularity. For these measurements, which relate to a very early state of the Universe, show that the Universe at that stage possessed a high degree of isotropy. The discovery is in agreement with recent theorems which demonstrate on theoretical grounds that in general relativity the effects of self-gravitation inevitably lead to isotropy and hence to the singular condition of the Universe in its initial state.

During the last few years we have been faced with this remarkable consequence, that both theory and observations indicate that ten billion years ago the material which now forms the Universe was in this immensely concentrated and high temperature state. Any attempt to visualize this early condition of the primeval concentrate in terrestrial terms leads to an impossible conceptual difficulty. The mass of the Earth is about 6×10^{21} tons. The Sun is a million times more massive than the Earth. The Milky Way is 100 billion times more massive than the Sun. Within the range of modern telescopes there are some 100 million galaxies. Thus the observable Universe contains more than 10^{46} tons of material. At these limits the number of galaxies observable continues to increase by two to three times for each fainter magnitude and hence the total mass of material in the primeval condensate must have been far greater than this 10^{46} tons.

In seeking a physical description of the early stages of the Universe it is first necessary to inquire how far back to the beginning our observations extend. We have seen that the observations of radio galaxies take us nearly halfway back to the beginning of the expansion. On the assumption that the quasar redshifts are wholly cosmological the look-back time is 80 to 90 percent—that is, to one or two billion years after the beginning of the expansion. At that stage most of the critical events in the history of the Universe had already occurred, for the present view is that the galaxies formed in a rather limited period of time, 100 million to a billion years after the beginning of the expansion. They condensed from a universe which at that epoch was probably a gaseous mixture of hydrogen and helium with a density about ten thousand times the present value.

The great importance of the discovery of the microwave background radiation lies in the fact that, if the current interpretation is correct, then these observations relate to a much earlier state of the Universe—to an epoch about a million years after the beginning. At that stage the temperature of the Universe is computed to have been about 5,000°K, the radius about one thousandth of the present value and the mean density a billion times greater than the present value. The photons which are today observed as the 2.7°K microwave background radiation are believed to have originated at this epoch and to have been propagating without interruption through space since that time. The difference in the calculated temperature of 5,000°K at that epoch from the 2.7°K observed today is compatible with a redshift of the radiation appropriate to the cosmological expansion of the Universe.

We believe, then, that the observation of the microwave background radiation provides us with a look-back time of 99.99 percent to the beginning of the expansion. At this stage when the temperature had fallen to a few thousand degrees, neutral hydrogen was able to form without being immediately photoionized by the surrounding radiation, and the radiation in the Universe no longer underwent substantial scattering or absorption. It was the epoch when the matter and the radiation in the early Universe ceased to interact freely. Before that time the Universe was dominated by radiation but when the temperature had fallen sufficiently for the neutral hydrogen to form, a decoupling of the radiation and matter occurred. Matter then emerged as the dominant component of the evolving Universe and after another 100 million to a billion years formed into galaxies by processes which are not yet fully understood. The photons which are now studied as the 2.7 degree microwave background radiation belong to the end of the radiation era, often called the primordial fireball of the Universe.

The study of this radiation helps to define the state of the Universe when it was about a million years old; at this early stage perhaps the most remarkable feature is that a feasible physical description can be given. The temperature was no greater than that associated with the surface temperature of an average star; the mean density was still very low compared with the densities of our terrestrial experience; and the electrons, protons, hydrogen, helium and photons are the familiar features of everyday physics.

The physical conditions in even earlier stages of the Universe, that is, in the first million years after the beginning of the expansion, can only be a matter for conjecture. As we approach closer to the beginning, the Universe must have been hotter and denser. Although we have no observational evidence of these early stages, it is important to enquire how much closer to the singularity, that is to time zero, the laws of physics and our knowledge of fundamental particles enable us to speculate. Can we formulate even in a speculative manner a possible sequence of events for the first million years after the beginning of the expansion to the epoch when the temperature had fallen to about 5,000°K and neutral hydrogen was able to exist and the photons measured today as the microwave background were emitted?

It is first necessary to emphasize that in the near vicinity of the singularity, at time zero, the physical theories known to us today fail completely. There is a point when we enter realms where classical concepts and quantum theory no longer suffice to describe the state of the Universe. In our ordinary macroscopic world it is possible to treat the gravitational field without encountering problems which arise from the discontinuities of quantum theory. However, as we consider smaller and smaller dimensions, a stage is reached where this situation is no longer acceptable. As the singularity is approached, quantum fluctuations of the gravitational field become so large that the laws of physics can no longer be valid. The dimension* at which this problem arises is 1.6×10^{-33} centimeter. In the expansion of the Universe from an imaginary infinite density of infinitely small size at zero time, this dimension would be reached 10^{-43} seconds after time zero, when the density would be 5×10^{93} grams per cubic centimeter. It is therefore possible only to speculate about the condition of the Universe when it was already 10^{-43} second old. The speculative calculations imply that at that point the temperature was in the neighborhood of 10^{33} deg K. The Universe might

---

*This "characteristic length" lg is given in terms of the gravitational constant G, Planck's constant $\hbar$, and the velocity of light c by the relation $lg = \sqrt{\dfrac{G\hbar}{c^3}} = 1.6 \times 10^{-33}$ cm. From the singularity $l = 0$, the corresponding time is $t_g \sim \dfrac{lg}{c} \sim 10^{-43}$ secs. The density is then $\rho g \sim \dfrac{c^5}{G^2\hbar} = \dfrac{\hbar}{cl^4} = 5 \times 10^{93}$ gms per cc.

then have been filled with short-lived exotic particles and anti-particles, about which we have little understanding, in equilibrium with the intense radiation field. The particles could interact to produce photons, and photons could interact to produce the particles. This conglomerate would have cooled very quickly as it expanded. After 10^{-4} second the temperature would have fallen to 10^{12} deg K and the density to 10^{14} grams per cubic centimeter. At this relatively cool stage the more familiar fundamental particles would have existed. The heavy particles and anti-particles of the earlier stage would have been annihilated and positrons, electrons and their anti-particles, together with some neutrons and protons, would have been in equilibrium with the radiation field.

A more realistic theoretical construction for the next phase of the expansion, from 10^{-4} seconds to 100 seconds after the beginning, seems possible. At 1 second the temperature would have decreased to 10^{10} deg K and the density to about 1 gram per cubic centimeter and electrons, positrons, neutrinos, anti-neutrinos, together with some protons and neutrons, would have been in equilibrium with the radiation field. At 100 seconds the temperature would have fallen to 10^9 deg K, but before this time it is projected that the main helium production in the Universe occurred.

Modern theories appear to give a satisfactory account of the formation of the other elements at a very much later stage in the history of the Universe, by thermonuclear processes in the interiors of stars. However, it appears impossible for these nucleosynthetic processes to produce a helium abundance of more than a few percent by mass, whereas today we find that astronomical objects possess a helium abundance of about 25 percent by mass. It is believed that this helium was produced in the period from 10 to 100 seconds after the beginning of the expansion, when neutrons combined with protons to form first deuterium and then helium nuclei.

After 100 seconds it seems probable that there was a long period extending to about a million years during which the temperature fell steadily, from 10^9 deg to about 5,000 deg K, as the Universe expanded. Throughout this period the helium nuclei, protons and electrons comprised an ionized gas in thermal equilibrium with the photons. At the end of this radiation era, or of the "primordial fireball" stage, the density had decreased to about 10^{-20} grams per cubic centimeter and, as we have already described, with the temperature falling to a few thousand degrees the hydrogen was no longer photoionized. The formation of neutral hydrogen marked the decoupling of matter and radiation. The microwave background photons observed today belong to this epoch. In principle therefore, at this age of about a million years the Universe entered the matter-dominated, observable phase. Although the occurrences in the Universe during the long period when the primeval hydrogen and helium formed into galaxies remain obscure, at least the problem lies in the possibly observable epochs of the history of the Universe.

THE FINAL STATE OF THE UNIVERSE

To a surprising extent we can look back to earlier epochs but we have no observational means of studying the future, although in so far as the present and future are determined by the past it is possible to predict the future in a speculative manner. We can be confident, for example, that tomorrow a measurement of the distant galaxies will show the same redshift as today, that the redshift of the radio galaxy in Boötes will still imply a recessional velocity of 46 percent of the velocity of light, and that the distance separating us from that object will have increased by about ninety thousand miles for every second of time. A separation of 324 million miles for every hour of time is not easy to contemplate in terrestrial terms. Compared with the scale of the Universe it is, however, a negligible change, since we are already separated from the object in question by more than 10^{22} miles. Even over the time span of a year the separation, which will have increased by over two thousand billion miles, represents a change of only one part in ten billion of our situation with respect to that galaxy.

Over the time intervals of terrestrial existence we can therefore make confident predictions about the future state of the Universe—it will appear unchanged when viewed on the large scale. But as we increase the time scale of prediction our confidence decreases. We are much less certain what the large scale condition of the Universe will be in the epoch 4 or 5 billion years from now when Earth becomes uninhabitable because the Sun will be exhausting its reserves of hydrogen. A further projection of ten times, to 40 or 50 billion years, takes us into an entirely speculative epoch. Only if the steady state theory had been found to be appropriate to the description of the Universe could we be confident. In that event the prediction would be that the Universe, on the large scale, would be the same as it is today. We believe that the Universe is not in a steady state but in an evolutionary condition, and it is therefore necessary to inquire what constraints can be placed on any description of the distant future.

In the various solutions of the equations of general relativity, in which the density of the Universe varies with time, two broad possibilities emerge for the future. Either the Universe will continue to expand indefinitely (the rate of expansion may decrease or tend to zero, but the Universe will remain open) or alternatively the expansion will cease and the Universe will collapse upon itself to another state of high density (the Universe will be closed). Physically the alternatives are determined by the mean density of the Universe. If the density of matter in the Universe is high enough, then the forces of gravitational attraction will ultimately overcome the forces driving the Universe apart and it will collapse.

It is possible to calculate the critical mean density at the present epoch. The figure is 2×10^{-29} grams per cubic centimeter. If the mean density exceeds this value, then there will be an ultimate collapse; if it does not, then the Universe will remain open. If we were aware of all possible forms of matter in the Universe we could calculate the mean density of the

Universe at the present time, and compare this with the critical value of 2×10^{-29} grams per cubic centimeter. The material in galaxies amounts to only a few percent of the critical density, but there is certainly more matter in the form of gas and radiation which might increase this value considerably. It is by no means impossible that future observations and a more detailed understanding of the various forms of radiations in the Universe and their origins—particularly X-rays— may lead to a more conclusive measurement.

At present it seems more probable that the answer may be obtained by measurement of the departure of the Hubble line from linearity at large redshifts. The curvature of the redshift-magnitude line for distant galaxies is a measure of the deceleration parameter, and this is directly affected by the amount of gravitating matter in the Universe. If the redshifts of the quasars are cosmological, then they refer to epochs when the effect of this parameter on the slope of the Hubble line connecting redshift and magnitude becomes marked. At present the large scatter in the results invalidates any conclusion. Nevertheless, this is, in principle, an observational parameter and it is reasonable to anticipate that the question as to whether the Universe is open or closed may be settled at some future time.

Until such a measurement can be made without ambiguity the final state of the Universe can only be described in terms of contrasting possibilities. If there is insufficient matter in the Universe to counteract the expansion, then more and more galaxies will move over the observational horizon, the energy of the Universe will become less and less available and the final state will be that of a heat death with the temperature of all matter approaching zero. On the contrary, if the density of the Universe is sufficiently great then the expansion will reach a maximum at some future epoch and the Universe will then contract to another singular condition of high density. The time scale for this reversal depends on the extent to which the density exceeds the critical value.

Some astronomers are confident that the constraints already placed on the slope and curvature of the Hubble line define the deceleration parameter so that the Universe will collapse. The most favored value for the parameter implies that we exist in an epoch at which about one tenth of the cycle has elapsed and that the turning point to a contracting Universe will be reached in another ten billion years. If, and when, this turning point occurs and the Universe begins to contract, the redshifts of the nearby galaxies will change to blueshifts and eventually distant objects will be observed as blueshifted. The time scale of contraction cannot be predicted—it is not clear whether the expanding and contracting phases must be symmetrical. Eventually the galaxies will merge, but the stars are unlikely to meet their end by collision. Before the density increases to the extent that collisions are frequent, it seems likely that the stars will be destroyed by external radiation, which will become hotter than the stellar interiors. Since the state of matter is unknown in the singular condition, it cannot be predicted whether the loop will be closed, or whether a "bounce" will occur, and the cycle of expansion and contraction repeat indefinitely.

VII

Space: Creation and Comprehension

The concept of space has exercised the minds of philosophers throughout the ages, and the logical arguments for and against the existence of a void have been common features of philosophical discussion. The logic of Parmenides, who lived in the fifth century B.C., was decisively against the existence of a void. When it was realized that what appeared to be a void actually contained air, the logic of the argument against the void was greatly substantiated. The argument is, indeed, as irrefutable today as it was two millennia ago. To those who maintain that something called empty space or a void exists, the response is quite clear, namely, if the void exists, then it is not-nothing and therefore is not the void.

The concepts of modern physics and cosmology have changed the nature and scope of the problem but the difficulty remains substantial. No one would agree today that the Universe which we observe and in which we exist contains a void at any time or place within itself. We know that even in the near vacuum between the galaxies, substance exists in the form of particles and radiation. The concepts of general relativity have raised these *internal* arguments to new dimensions of thought. Nevertheless there is a most formidable problem if we believe that modern observations imply a singular, or near singular, origin of the Universe. We have seen that the consequences of the existence of the microwave background radiation are accepted as indicative of a high temperature condensate in the early stages of the evolution of the Universe. This primeval matter was concentrated, and its mass, density and size as time evolves from the zero of time, when the expansion began, can be specified in conventional language. In the light of contemporary knowledge one is bound to inquire whether this condensate—and indeed the Universe as it exists today—had an existence in an infinite void which contained nothing, or alternatively whether the entire concept is meaningless because time and space may have no meaning apart from the nature which *we* attribute to them, hence these concepts suffer the limitations of human thought. In the former case we have to acknowledge that the logical argument of Parmenides remains as powerful today as it did over two millennia ago. In the latter case we accept, not merely a limitation of human thought, but a most remarkable and unsuspected entwinement of man with the actual existence of the Universe and of time and space. The conventional response of the scientist faced with this dilemma is to escape with the remark that the question does not lie within the realms of scientific inquiry. The response is no longer valid. Our observations during the last decade make it no longer possible to escape with obscure notions of infinity, and contemporary theory deals with the first seconds of a universe which, in terms of human measurements, is conceivable as a localized entity.

SPACE IN NEWTONIAN THEORY AND GENERAL RELATIVITY

These problems are of recent origin because we no longer have alternative avenues of escape from the consequences of our observation of the Universe. It is the near compatibility of our cosmological measurements with the general theory of relativity which has undermined the long held conception of an absolute space in which matter could exist *independently* of the infinite space which also existed. It is a remarkable feature of scientific history that, notwithstanding the force of the logical argument of Parmenides, the belief in absolute space and absolute motion survived until Einstein published his general theory in 1917. The reason is that, as in so many other fields, physics was dominated by the reasoning of Aristotle. He insisted on the distinction between matter and space, and his opinion that space was a receptacle in which matter might or might not exist remained essentially the scientific view for two thousand years. The critical sentence occurs in his *Physics:* "The theory that the void exists involves the existence of place: for one would define void as place bereft of body." Although undermined by Einstein's theory in 1917, it remains, of course, the commonsense view today.

The Aristotelian view was implicit in the scientific revolution of the sixteenth century, where because of the most elementary considerations, concerning sunrise and sunset for example, the heliocentric motions had to be regarded as absolute. In the seventeenth century the argument became explicit in the Newtonian theory. Without ambiguity Newton asserted the existence of absolute space and of absolute motion in that space. For him the properties of a body were independent of the space in which it existed and the motion of a body through this space was an absolute motion. In 1717, he wrote in his *Opticks:* "All these things being consider'd, it seems probable to me, that God in the Beginning, form'd Matter in solid, massy, hard, impenetrable, moveable Particles, of such Sizes and Figures, and with such other Properties, and in such Proportion to Space, as most conduced to the End for which he form'd them; . . ." and as regards their motion, "It seems to me farther, that these Particles had not only a *Vis inertiae,* accompanied with such passive Laws of Motion as naturally results from that Force, but also that they are moved by certain active Principles, such as is that of Gravity, . . ." With an infinite universe existing in an infinite space which also existed, Newtonian theory seemed logical and complete, and remained so for another two and a half centuries.

Although Newton's view of space prevailed it was not uncontested. His dispute with Leibniz extended to the nature of space as well as the infinitesimal calculus. In maintaining that space was not real but had a real counterpart in the arrangement of the monads (the term Leibniz used for "substances") in a three-dimensional order, Leibniz, it could be said, was closer to the modern view than was Newton. Half a century later Kant, in *The Critique of Pure Reason,* evolved a complex argument for the subjective and transcendental nature of space which seems to have a particular relevance to the formidable conceptual problem which we face today in the contemplation of the Universe.

Throughout the history of astronomy there are many examples in which a correct philosophical argument has lain dormant until it has been possible to cast it as a mathematical formulation, or prove the argument by observation. Whereas in the *Principia* Newton wrote that space was absolute which "in its own nature, without regard to anything external, remains always similar and immovable," Bishop Berkeley in his *De Motu,* published in 1721, maintained the precise contrary: "every place is relative, every motion relative. If all bodies are destroyed we shall be left with mere nothing, for all the attributes assigned to empty space are immediately seen to be privative or negative except its extension. But this, when space is literally empty, cannot be described or measured and so it too is effectively nothing." Towards the end of the nineteenth century, 150 years after the publication of *De Motu,* the Austrian physicist and philosopher Ernst Mach gave a more specific connotation to Berkeley's argument. In *Science of Mechanics,* Mach stated his view that the behavior of any one body in the Universe could only be considered in relation to all other bodies in the Universe: "When we say that a body preserves unchanged its direction and velocity *in space,* our assertion is nothing more or less than an abbreviated reference to the *entire universe.* "

The final formulation of the mathematical argument was published by Einstein as the general theory of relativity. Einstein generalized the principle enunciated by Mach, that the inertial properties of matter on a small scale are determined by the behavior of matter on a cosmic scale. He referred to the concept as *Mach's principle* and the universal relationship of inertial masses is commonly known by that name.

This problem of the distinction between gravitational and inertial mass had existed since Newton formulated his laws of gravitation. He assigned to each body a gravitational mass and it is to this that we refer when calculating the force of attraction between two bodies. In addition, the second law of motion states that the acceleration of a body depends on the force applied to it, and on another property which Newton called the "quantity of matter" possessed by the body. Today we refer to this as the inertial mass of the body, the inertia of the body being its resistance to motion. In the formulation of the Newtonian laws these two properties of the body are quite distinct—the gravitational mass determines the force exerted on another body, whereas the inertial mass is a measure of the resistance of the body to motion. The functions of the two properties appear to be entirely different, yet Newton could not measure any difference between the two quantities by experiment, neither could any difference be detected in many subsequent and more delicate measurements. In 1890, for example, Roland V. Eötvös proved that the inertial and gravitational masses of a body were equal to 1 part in 100 million. Classical physics cannot account for this but in general relativity the identity appears as a natural consequence of the theory because gravitational forces are manifestations of the inertia of bodies in space, which is itself modified by the presence of these bodies.

Although in our normal life Newtonian theory appears as a workable and satisfactory approximation of Einstein's theory there are vital differences separating the two concepts. Furthermore, as we have seen, in various astronomical tests of the two theories, the evidence is conclusively in favor of general relativity. The Newtonian concept of an absolute space and absolute motion must therefore be abandoned. For Einstein, the properties of space are a function of the bodies contained in the Universe. A ray of light grazing the Sun's disk is deflected because space-time is curved in the neighborhood of the Sun. The practical effect can be calculated from the theory and measured observationally. In fact the modern application of Newtonian theory also predicts that the light will be deflected—but for a different reason. Because of the equivalence of mass and energy implied by the *special* theory of relativity, a ray of light assumes the property of inertial mass, and the Newtonian laws lead to the prediction that the ray will be deflected by nearly one second of arc if it grazes the Sun on the way to an observer on Earth. The calculated deflection in Einstein's theory is twice that of classical theory, and the observed effect is a decisive rejection of classical theory.

The failure of the Newtonian concepts and the applicability of the general theory to the observed Universe represents a decisive rejection of the commonsense view that on the one hand we have a body, and on the other hand we have a space which is a receptacle for that body. The philosophical problem is reminiscent of Leibniz's argument that space is not real but merely has a real counterpart in the arrangement of the monads. In modern phraseology, we note that space is curved in the vicinity of a massive body. Does the massive body cause the curvature of space-time, or is the curvature itself responsible for the existence of matter?

This philosophical query emerges as a most formidable one when we consider the implications of general relativity for the problem of the beginning. Although the theory leads to a variety of possible evolutionary models, all have in common the difficulty implied by the singularity of infinite density and zero radius of the Universe at the beginning of the expansion. As we retrace the history of the Universe towards time zero, we conclude that matter increases in density and the space curvatures reach extreme values. At 10^{-43} seconds from the beginning, before the space curvature becomes infinitely great, the language of modern physics loses its meaning and we begin to feel the force of the metaphysical argument of Kant, that it is possible to imagine nothing in space, but impossible to imagine no space.

CREATION

Throughout the history of the human race there have been beliefs about creation and the beginning. The earliest records of such ideas emerge in the ancient states of Egypt and Mesopotamia. The hieroglyphic texts on the pyramids of the pharaohs of the fifth and sixth Dynasties (2480–2137 B.C.) reveal a conception of a primordial state before the physical Universe existed. In one case, the Pharaoh Pepi is assured that he was engendered by the god Atum when "existed not the heaven; existed not the Earth; existed not men; before the birth of the gods; before the existence of death." Another declaration of that epoch proclaims that the king was "born in Nun, when heaven existed not, and Earth existed not, when existed not that which was to be established, when (the) disorder existed not." "Nun" was the primeval waste of waters and it is evident from these texts that the Egyptians envisaged the primordial state as a watery chaos or waste.[1]

For the last two millenia the Biblical story of the Creation has had a powerful effect on Western culture and imagination. The ancient texts imply a creation out of *something*. In the thirteenth century A.D. Saint Thomas Aquinas refuted this as being an error against God. In Book II of *Summa Contra Gentiles* he insists that God created the World out of *nothing*. In this he supported Saint Augustine, who centuries earlier (in *Confessions*, Book XI) had maintained that Creation occurred as asserted in *Genesis*, that is, out of nothing. The schools of Saint Augustine and Saint Thomas Aquinas sought to maintain the belief in Creation by God out of nothing, as opposed to the views of Plato and Aristotle, who believed God to be an architect giving form to primitive matter, rather than a Creator.

In the seventeenth century the development of science

began to exert powerful influences on these views of creation. Newton believed, as we have seen, that God "form'd Matter in solid, massy, hard, impenetrable, moveable Particles" and "no ordinary Power being able to divide what God himself made one in the first Creation." The Newtonian laws provided an extremely powerful intellectual tool leading to the development of a cosmology which seemed capable of explanation by the physical laws of force and matter.

In the twentieth century considerations of Creation and of a beginning have been unfashionable. Furthermore, science has become a powerful experimental and observational activity to the extent that although many issues remain without explanation the belief has emerged that the outstanding problems will yield to scientific inquiry in the future. Here we are not concerned with the generality of this issue but solely with the contemporary scientific outlook on the problem of creation near time zero, and the expression of this problem in the modern language of physics and astronomy.

Today it is extremely doubtful that any cosmologist would claim the certainty of earlier beliefs when faced with this question, although in terms of nuclear physics, quantum theory and general relativity it is possible to construct a theoretical model for the early history of the Universe within certain limitations. These limitations are based on the observation of the microwave radiation, which we believe refers to the time a million years after the beginning of the expansion, and on a theoretical extrapolation to earlier moments in terms of the temperatures, densities and nuclear reactions which could give rise to this observed condition. This backward extrapolation becomes less credible as time zero is approached. At the extremely high temperatures and great densities involved in the first milliseconds of the expansion it is presumed that heavy fundamental particles were in a short-lived equilibrium with the intense radiation field. The process of production of the particles from the radiation field and of the photons by particle interaction are understood in principle. Since in the earlier and higher temperature stages of this period the photons would have the highest energies, the most massive particles—the baryons and anti-baryons—would be present. At temperatures in the range 10^{12} to 10^{33} deg K and at the densities appropriate to those epochs, no real understanding exists about the interaction or behavior of these particles. Nevertheless the concepts of modern physics seem applicable to that stage in the expansion.

Greater difficulties arise as the extrapolation continues. Earlier than 10^{-23} seconds after the beginning it is believed that the space curvature would be so great that pairs of particles would be created from the intense gravitational field (in a manner analogous to the creation of electron-positron pairs in strong electromagnetic fields). At 10^{-43} seconds the space curvature reaches the extreme value conceivable in terms of general relativity and quantum theory. At this stage we reach the barriers of physical theory, where further understanding, if possible in any physical sense, must depend on a quantization of the gravitational field—a unification of fundamental theory which has so far proved to be beyond the human intellect. For in dealing with the behavior of matter and radiation under these extreme conditions scientists reach a curious confrontation with the fundamentals of gravity and particle physics.

General relativity is concerned with gravitational forces. It is these gravitational forces, familiar in our everyday lives, and the theory of space-time associated with these forces, which determine the large scale nature of the Universe, and lead to the concept of a universe evolving from a singular condition. On the other hand, when we probe into the physical structure of the Universe, we deal with atoms and their constituent particles, where the predominant forces are not those of gravity. Three other types of force are recognized. First, there is the electromagnetic interaction between particles which have an electric charge—like charges repel and unlike charges attract. This electromagnetic interaction between two particles is millions of times stronger than the gravitational force. Electromagnetic forces, like gravitational forces, are well known in the macroscopic world. In the nucleus, however, quite different forces must predominate, because protons, for example, are held together strongly in spite of the electromagnetic repulsion between them. Two types of nuclear force are recognized—the strong nuclear force, and the weak nuclear force. It is the strong nuclear force which binds protons and neutrons together in the atom. In the atomic nucleus this force is 10^{40} times stronger than the gravitational force; it is independent of charge and its influence extends only over distances comparable with nuclear dimensions. Two protons in a nucleus, for example, having the same charge suffer an electrostatic *repulsion* (equivalent in strength to the gravitational pull of the Earth on a mass of 25 kg at the surface), but when the distance between the protons is less than 10^{-13} cm, the nuclear *attractive* force exceeds this repulsive force. The weak nuclear force operates in certain processes where transformation of particles occurs—as in β decay—and is much weaker than the electromagnetic interaction. These nuclear forces are explained in terms of exchanges between fundamental particles, whereas the gravitational forces do not, as yet, have a physical explanation apart from the hypothetical gravitons.

It is the reconciliation and the integration of these various forces into a coherent theory which presents a fundamental problem of such difficulty that there are contemporary thinkers who suggest that the real structure of space and time may lie completely outside our present concepts, or even that the very foundations of mathematics and its logical axioms are insecure.

Modern physics and astronomy have transformed the problem of creation to new and erudite concepts. Particles may be created from intense fields of force. In the realms of the beginning which lie beyond the present descriptive limits of physical theory this may be the only type of scientific conclusion which is possible. We may well feel like the poet Hesiod that "Verily at first Chaos came to be. . . ."[2]

COMPREHENSION

It is a strange feature of our lives that we exist in an age when our scientific knowledge is immensely greater than that possessed by any previous generation, and yet this knowledge has destroyed the fabric of our understanding about the fundamental cosmological problem. In that previous age of scientific advance which also led to the destruction of long held views—the Copernican-Galilean era, before the Newtonian synthesis—Francis Bacon defined an attitude which might well commend itself to us today. In *The Great Instauration* he wrote that no one need be alarmed at a suspension of judgment "in one who maintains not simply that nothing can be known, but only that nothing can be known except in a certain course and way; and yet establishes provisionally certain degrees of assurance, for use and relief until the mind shall arrive at a knowledge of causes in which it can rest." So it has been throughout the ages. In the sixteenth century two men, separated by a thousand miles, simultaneously imprinted their certainty of knowledge on the problem of the Universe. On the ceiling of the Sistine Chapel, Michelangelo translated visually the drama of the Creation. In these great frescoes of the *Creation of Man* and in the surviving copies of the Copernican *De revolutionibus orbium caelestium* we have examples, which are perhaps unsurpassed, of Man's attempt to comprehend the Universe.

We do not know what legacy of comprehension the twentieth century will bequeath to mankind. In the first decades of the century, quantum theory and general relativity seemed for a time to remove the obstacles to understanding which had emerged in Newtonian mechanics and classical theory. We reached a stage when it appeared that Planck and Einstein had achieved a synthesis for new knowledge similar to that achieved by Newton over two hundred years earlier. Now, as we enter the last decades of the century, we see that this is not the case. Quantum theory and relativity form a partial synthesis only. The progressive penetration into the remote regions of time and space, and into the innermost structure of the atom have simultaneously revealed that neither theory enables us to comprehend the Universe. A further step is required either to synthesize the theories or to supersede them by new concepts which could well be as revolutionary to our outlook on the physical world as the iconoclastic ideas of Planck and Einstein.

At the same time we should reflect on the vast extension of our comprehension of the external world made possible by Planck and Einstein. During the last few years experiments have been carried out in the high energy physics laboratory of Stanford University using the particle accelerators to study the interaction of electrons and positrons at high energies. Features of these collisions, such as the angular distribution of the particles and the momentum transfers in the interactions, can be calculated on the basis of quantum theory. The theory has been found to be correct down to distances of about 4×10^{-15} centimeters. We know also that at the opposite extreme the quantum theory is applicable to processes far out in the environment of the Earth, in the magnetosphere. Thus quantum theory is known to be applicable to electrodynamics and particle physics over a distance ratio of at least 10^{25}. The predictions of general relativity have been verified over cosmological distances and, if we accept the implication of the microwave background, over a time scale of nearly ten billion years, representing 99.99 percent of the predicted history of the Universe. These are remarkable achievements, but we do not yet comprehend the relation between the two theories, and because of this we cannot resolve the conflict which emerges when we inquire about their ultimate limits of application. Notwithstanding the cosmical scope of general relativity, there is a built-in failure of the theory determined by three constants of nature, the gravitational constant G, Planck's constant $\hbar$, and the velocity of light c. These define a length of 10^{-33} centimeters beyond which no progress can be made unless the unification of the two theories is possible. Such a unification may not be possible, and further comprehension may demand radically different concepts leading to a new view of nature which embraces quantum theory and relativity in the same way that these theories embraced Newtonian gravitation and classical electrodynamic theory.

Many strange and bewildering features of the physical world emerge at these limits which seem incomprehensible in terms of our present outlook. There are strange properties of nature at the Planck limit of 10^{-33} centimeters. It is possible to calculate the energy and the mass of a quantum possessing that wavelength. Such particles would have the peculiar property that the gravitational attraction between them would overwhelm all other forces instead of being about 10^{40} times weaker as in the world of our experience. The natural world which we inhabit and investigate and to which quantum theory and relativity apply is a known world. The transference to the unknown world where physical theories fail leads to the search for the new synthesis or new theory—and to the more fundamental query about the *possibility* of attaining such knowledge.

The transference may lie beyond scientific comprehension. It is not clear whether the difficulties arise because we have externalized the object of investigation. If this is the case then it may be asked whether there can be any reality in an externalized procedure for the investigation of an entity of which we are a part. Of course, this is an ancient problem in the theory of knowledge, but here we have a special circumstance: physical theory has defined a localized entity in time and space which, although capable of localization in human thought, is also the entirety of time and space. In the known world we achieve scientific knowledge because we can proceed on the assumption that the object of investigation exists independently of us, even though the precision of the knowledge is ultimately limited by the operation of the uncertainty principle. In seeking knowledge about the state of the Universe near time zero, however, it is not evident that such externalization is possible. The issue is deeply involved in the interpretation of quantum theory, since the question is whether there can be a consistent interpretation of quantum theory without external concepts, or without the interaction of living or conscious beings with the object of the investigation.

The immediate tendency to answer this in the negative is disputed by some contemporary thinkers. One line of argument in favor of consistency denies that the appropriate wave functions are ever converted into knowledge in the sense of a classical probability.[3] However, it is difficult to concede that these considerations can be applied where statistical knowledge is not possible—and in the ultimate case we are concerned with a single unique entity, the initial state of the Universe.

There are, indeed, profound problems concerned with the presence of man in the Universe which envelop the localized problem of the emergence of life on Earth discussed in Chapter 4. We have attempted to describe the early stages of the expansion of the Universe but the *description* in terms of nuclear physics and relativity is not an *explanation* of those conditions. Formidable questions arise and it is not clear today where the answers should be sought; indeed, even the scientific description of these queries produces the remarkable idea that there may not be a solution in the language of science. Why is the Universe expanding? Furthermore, why is it expanding at so near the critical rate to prevent its collapse?

The query is most important because minor differences near time zero would have made human existence impossible. When the Universe was one second from the beginning of the expansion we have stated that the temperature had fallen to 10^{10} deg K and the density to 1 gram per cubic centimeter. It is a phase when, it is postulated, the Universe had already reached the possibility of description in terms of common physical concepts. If at that moment the rate of expansion had been reduced by only one part in a thousand billion, then the Universe would have collapsed after a few million years, near the end of the epoch we now recognize as the radiation era, or the primordial fireball, before matter and radiation had become decoupled. This remarkable fact was pointed out recently by one of the most distinguished contemporary cosmologists[4] who referred to the suggestions[5] that out of all the possible universes the only one which can exist, in the sense that it can be known, is simply the one which satisfies the narrow conditions necessary for the development of intelligent life.

The essence of our presence in the Universe today is that *we* require the Universe to have certain properties. Long before we reach the problems of biological evolution on Earth three to four billion years ago, we face this more fundamental question. At least one essential condition of our existence is that the Universe must expand at almost precisely the rate at which we measure it to be expanding. If the rate had been less by an almost insignificant amount in the first second, then the Universe would have collapsed long before any biological evolution could have taken place. Conversely, if the rate had been marginally greater, then the expansion would have reached such magnitudes that no gravitationally bound systems (that is, galaxies and stars) could have formed.

The problem is related to the observations of the microwave background radiation, which reveal the surprising fact that the Universe is isotropic to at least one tenth of a percent. We do not know why the Universe is so evidently anisotropic locally (galaxies, clusters of galaxies), whereas on the large scale it is so nearly isotropic. This extraordinary large-scale uniformity seems to imply that, after all, the Universe is nearly spherically symmetric about *us*, and that the condition of isotropy is closely connected with the rate of expansion which we measure—a rate so delicately involved with the possibility of our emergence in the Universe.

The circumstances of isotropy and the rate of expansion are not the only features of the early Universe which were critical to our existence today. We have referred to the formation of helium in the epoch ten to one hundred seconds after time zero. We believe that most of the helium in the Universe today was formed at that early stage. The processes of formation from neutrons, protons to deuterons and then to helium are well understood. But why were not all the protons in the Universe absorbed in this process, leaving either deuterons or helium nuclei as the major elements? If this had happened, then, after a million years, when the temperature had fallen to a few thousand degrees and photoionization no longer occurred, the Universe would have contained deuterium or helium. There would be little hydrogen. The result would have been catastrophic, because the thermonuclear processes in the Sun (and the stars) on which the emergence of intelligent life depends so critically, is based on hydrogen and not deuterium. The difference is between the steady output of energy over billions of years as in the Sun, and the explosive reaction of the hydrogen bomb—between the strong and weak nuclear forces when particles approach to distances of about 10^{-13} centimeter. In the Sun the basic transformation involves the fusion of protons. In the bomb the process involves the fusion of deuterons. The proton + proton reaction is weak; the deuteron + deuteron is a strong nuclear interaction. The difference in the speed of the reactions is immense —about a billion billion times. Thus if deuterons, or deuterium, had been the major product of the early radiation era of the Universe then the stars would have exploded as soon as they began to form. The existence of protons (and their subsequent recombination with electrons after a million years to form hydrogen), instead of deuterons, was an essential factor in the subsequent evolution of a universe in which life could emerge. On the other hand, if the nuclear reactions in the first hundred seconds had converted all the protons and deuterons into helium nuclei (instead of about 25 percent), then no long-lived stars like the Sun could have been formed, and, again, no life could have emerged.

The backward projection of the physical content of the Universe implies that about a millisecond after time zero (at the end of the heavy particle, baryon, era described in Chapter 6), the numbers of protons and neutrons were about equal. The neutrons decayed into protons and electrons with a half-life of twelve minutes, and the number of protons increased at the expense of the neutrons. By the time the Universe was one hundred seconds old the critical ratio of protons to deuterons to helium nuclei had already been determined.

Helium originated at that epoch through the sequential process of the neutron + proton reaction to give deuterons and then helium. But the interaction of two protons can form the helium-2 isotope, which decays spontaneously into deuterons. Hence once more we have to inquire why any protons were left to form hydrogen? Quite apart from the era of helium formation in the first hundred seconds, why did not all the remaining protons interact to give helium-2? The answer is that the helium-2 state is unstable, but only marginally unstable—by half a million volts. Since the proton + proton attractive force is about 20 million volts, the difference is only about 2.5 percent. In other words, if the proton + proton reaction had been only a few percent stronger, then all the protons would have formed into helium in the first few million years of the Universe, and, once more, a universe would have evolved which could never be comprehended by any form of intelligent life.

It seems that the chances of the existence of man on Earth today, or of intelligent life anywhere in the Universe, *are* vanishingly small. Is the Universe as it is, because it was necessary for the existence of man? Is there a false logic in the argument, or are the basic axioms of our mathematics and physics wrong? We start from the knowledge of our existence on Earth. Over several millennia, and especially over the last four hundred years of Earth time, we have established a scientific knowledge of the natural world which enables us to comprehend the Universe on a cosmical scale, that is, over scales of time and distance which are difficult but not impossible to understand in terrestrial terms. Further, the nature of our comprehension enables us to test the validity of our physico-mathematical concepts on this cosmic scale. In this way our comprehension established in the language of modern science has been tested for 99.99 percent of the age of the Universe. Of course, there are many details to be filled in, but the broad scope of the cosmos is explicable within this framework, and although we have no scientific observations directly related to the first million years, it is possible to give a plausible scientific description of the evolving Universe during this early period. We reach two conclusions. The first is that the processes of comprehension lead to an exotic, almost bizarre, circumstance near time zero, where neither logic nor physical concepts, but only the infinities and infinitesimals of mathematics apply. To explain that the Universe "began with a hot big bang" is a nonsensical colloquialism. It is, as Newton wrote nearly three hundred years ago, "an occult specifick Quality" telling us nothing. The second conclusion is that our comprehension leads to a scientific description of an evolving and expanding universe. The mathematics and the physics enclose a variety of such possible expanding universes, but our *measurements* narrowly define one such Universe—which had to be that particular Universe if it was ever to be known and comprehended by an intelligent being.

These seem to be the extraordinary and impossible conclusions of contemporary scientific investigation of the Universe, and it is hardly surprising that many eminent mathematicians and cosmologists are today examining the fundamental mathematical and physical assumptions on which the projections are based. The theory of continuous creation, or the theory of the steady state, was the result of one such examination, but the observational evidence has subsequently been found to be at variance with the predictions of the theory. Since the acceptance of the evidence that the Universe has evolved from a high-temperature initial condition, a great amount of work has been devoted to possible theoretical constructions whereby the singularity at time zero is avoided—for example, by transference to a previous cyclic phase of the Universe before the quantization of the gravitational field becomes important (say, at times less than 10^{-23} second).[6]

Suggestions have also been made that the fundamental constants of nature had different values in earlier times. Forty years ago P. A. M. Dirac drew attention to the occurrence of certain large numbers. For example, it is possible to express the age of the Universe in units of time that are fixed by the constants of atomic physics (that is, the electronic charge and mass, Planck's constant, and the velocity of light). A very large number, 10^{40}—which happens to be the same as the ratio of the electric to the gravitational forces between two charged particles—is obtained. The square of this number, 10^{80}, is equal to the number of nucleons in the Universe (assuming the density of matter is the critical density separating the open and closed universes). Dirac's theory is that such large numbers vary with the age of the Universe. A consequence of the theory is that the gravitational constant G should vary inversely as the age of the Universe. In order to reconcile this with general relativity, which presupposes the constancy of G, it is necessary that the number of nucleons should increase with time. This leads to a form of continuous creation cosmology which embraces the concept of a universe evolving from a dense initial state. It is a cosmology in which each atom has a probability of reproducing one of its kind, and leads to peculiar predictions—for example, that a piece of rock should grow in size over geological ages, and that the Earth-Moon distance should increase with cosmological time, at the rate of about two centimeters per year. Another consequence of the theory is that the redshift can be explained without recourse to the expansion of the Universe—the wavelength observed is simply related to the epoch when the light was emitted, and there is no need to invoke a recession of the galaxies. The theory predicts that the luminosity of a star should actually increase with time, that the Sun will get hotter as time passes. In the time scales of the theory the Universe has always existed. The implied rate of change of G is only a few parts in 100 billion per year. Since the most accurate measurement of the rate of change of G can only set limits about ten thousand times more than this predicted value, we do not yet know whether Dirac's theory will contribute significantly to solving the cosmological problem. If improvements in the techniques of measurement enable the predictions to be tested, and if they are found to be justified, then man's problem of comprehension would be translated from contemplation of the infinities at the beginning to that of the infinite duration of time.

One of the most eminent cosmologists working today, academician Ya. B. Zel'dovich, delivered the inaugural address at the cosmological symposium in Poland in 1973, which was arranged by the International Astronomical Union in celebration of the five-hundredth anniversary of the birth of Copernicus. He posed the question "What was the state of the Universe in the remote past? Was it a well ordered expansion with only small departures from strict-uniformity and isotropy? Or perhaps it was strongly turbulent and chaotic. . . . How do we make the choice?" In answering he remarked that the scientific investigation of these questions was based to an equal extent on logic and intuition, on hypothesis and rigorous proof. He said that an objection to this procedure was that it depended "on individual prejudices, the likes and dislikes of authors, and perhaps even . . . on their subconscious Freudian attitude to such things as order, chaos, anti-matter."[7] The theories, the doubts and prejudices fade only when there can be a confrontation with measurement and observation. Today it seems unlikely that there can ever be such a confrontation within the epochs of space-time where the possibility of man's existence in the Universe was determined. We are reminded of the scepticism of the eighteenth-century philosopher Hume, who wrote in his *Treatise of Human Nature* that "all our reasonings concerning causes and effects are derived from nothing but custom; and . . . belief is more properly an act of the sensitive, than of the cogitative part of our natures."

VIII

The Confrontation Between Man's Creative and Destructive Activities

In the American weekly publication *Soviet Aerospace*[1] of 3 November 1975, there was an account of the successful landing on Venus of the Soviet spacecraft Venera 10 on 25 October, together with reproductions of the photographs of the surface of the planet transmitted by Venera 9 on 22 October 1975. This account of one of the spectacular triumphs of modern science was preceded by references to the estimates of the Central Intelligence Agency that the expenditure on defense in the Soviet Union in 1974 was over $93 billion, or a fifth higher than the $70 billion spent by the USA—and that this probably amounted to 15 percent of the gross national product compared with about 7 percent in the USA. In recent years it was estimated that the Soviet Union had produced twenty designs of Intercontinental Ballistic Missiles compared with seven in the United States. The same publication on 4 October 1976 carried an account of statements made by Defense Secretary Donald Rumsfeld that the Soviet Union was able to deploy forty new versions of an ICBM, each carrying a nuclear warhead of twenty to thirty megatons. Since the Strategic Arms Limitation Talks between the USA and USSR then in progress (SALT II) proposed a limit of 2,400 delivery systems and 1,320 MIRV (Multiple Independently Targeted Reentry Vehicles) missile launchers, it is evident that both East and West are equipped with destructive powers of frightening immensity. This linkage, in a few pages of a Western weekly publication devoted to Soviet aerospace, of the successes of space science with the destructive implements of mankind epitomizes a frightening dilemma of the contemporary world. The unpleasant fact is that the activities in space science, which make possible our exploration of the planetary system, are themselves made possible by the military development of the rocket. The rockets which launch the space probes to Venus and Mars may differ in detail but do not differ in principle and power from those poised to deliver the hydrogen bomb. There is, of course, a long history of attempts to use some form of rocket weapon in warfare. In A.D. 1232, thirty thousand Mongols led by Ogotai Khan, the son and successor of Genghiz Khan, attacked the northern Chinese capital of Piang-King (K-'ai-fêng-foo). The Chinese defenders met the onslaught with powder-propelled fire arrows—a new and devastating weapon of that day which made a noise like thunder and traveled five leagues; where it fell, a fire extended for two thousand feet. From that period there are frequent references to the use of rockets in warfare, but although there are many examples of their successful use in battles, they were too inaccurate and unreliable to be regarded as a major weapon. For example, six centuries after the battle of Piang-King, the most advanced rockets of the day were used in Wellington's campaigns, but he had a poor impression of them. An eyewitness of his 1814 campaign wrote that the rocket was "a very uncertain weapon. It may indeed spread havoc among the enemy, but it may also turn back upon the people who use it, causing, like the elephant of other days, the defeat of those whom it was designed to protect."[2] The use of the rocket as a weapon declined thereafter. Even its usefulness as a signaling device decreased after the advent of radio in the early twentieth century. As a means of attack it could not compete with modern developments in conventional artillery. The subsequent progress culminating in the conflict of interests today, provides a classic example of the interaction of science and war.

At the turn of the century there were three pioneers who began the line of development resulting in the ballistic rocket. They were K. Tsiolkowsky in Russia, R. H. Goddard in America, and Hermann Oberth in Germany. They worked independently and they became interested in the subject because they believed that rockets held the key to flight in space. Tsiolkowsky was a theorist; it was he who obtained the correct theoretical solutions and described the principle of the multistage rocket. Goddard succeeded in launching the world's first liquid-fueled rocket from a farm in Massachusetts in 1926. It rose 4 feet, traveled 184 feet and fell to Earth after 2.5 seconds. Although Tsiolkowsky and Goddard made immense theoretical and practical contributions to the problem of rocket flight, their work was effectively disregarded by the Soviet and American governments respectively until after World War II. (In 1960 the United States government made a settlement of a million dollars on Goddard's widow and the Guggenheim Foundation for rights to two hundred of Goddard's patents—"basic inventions in the field of rockets, guided missiles and space exploration.")

THE GENESIS OF THE V2 ROCKET

On 6 September 1944, the Chiefs of Staff of the Allied armies announced that all the launching sites of the German V1 missile (the buzz bomb) had been captured. Since June of that year more than eight thousand of these low-flying missiles had been launched against London. They had killed over six thousand people and injured another eighteen thousand. The relief that these attacks had ended was short-lived. Three days later, on 9 September 1944, the Germans launched the first ballistic rocket—the V2—against London. There was no effective countermeasure, since the rocket attained an altitude of 60 miles and reached its target, 200 miles distant, at a speed of 3,500 miles per hour, 5 minutes after liftoff. It was fortunate for the Allies that Hitler had been so tardy in recognizing the potential of the V2. For although the allied offensive had captured all the launching sites by 27 March 1945, 1,115 of these weapons had fallen on London and southern England, killing 2,754 and injuring 6,523 people.

By strange twists of fate, the V2 owed nothing to Tsiolkowsky or to Goddard. In 1923 the German theorist Hermann Oberth—who in the 1960s was to watch the liftoff of an Apollo moon rocket from Cape Kennedy—published independent solutions of the major problems of rocket flight. He became president of the German Society for Space Travel in 1929. This powerful society carried out several successful tests in some of which the rockets reached altitudes of a few thousand feet. However, after the successes of 1931, the society's affairs declined sharply because of the economic depression, and in desperation they sought the help of the German army. The most significant result of this move was that the German army invited a young student, Wernher von Braun, who had joined the German Rocket Society in 1929, to complete the work on his doctoral thesis at the army proving grounds at Kummersdorf. (The international ties of the society were intolerable to Hitler and the Gestapo and in 1934 their testing ground was turned into an ammunition dump.)

In this manner the vision of rockets for space flight became a vision of rockets for destructive purposes. First the fear of war and then war itself stimulated the German army rocket group under Captain (later General) Walter Dornberger to develop the rocket on the basis of twentieth-century science and technology. Von Braun became technical director of the group, and in April 1937 he moved with a staff of eighty from Kummersdorf to Peenemünde on the Baltic Coast. There, modern rocket technology was created. At last Hitler was impressed, and in 1943 he ordered that top priority should be given to the work at Peenemünde and to the development of the V1 missile and the V2—the world's first ballistic missile. The great difficulties and political intrigues surrounding these developments have been described by von Braun.[3] The duplicity of the situation even in war is illustrated by von Braun's account of his encounter in February 1944 with Himmler, who tried to coerce von Braun into deserting the army and working for him. After refusing, von Braun was arrested by the Gestapo and accused of working for space exploration and not the war. The intervention of Dornberger with Hitler secured his release and the ultimate success of the V2 rocket.

THE PATH TO SPUTNIK 1

Twelve years after the end of World War II, the first artificial Earth satellite, Sputnik 1, was launched from the Soviet Union on 4 October 1957. Historians will reflect on that episode as one of the great turning points in the history of the twentieth century. Sputnik was acclaimed by the world as a major enterprise which suddenly opened a vast new era for the scientific study of the planetary system and the Universe. Twenty years later, it is readily seen that this claim was entirely justified. However, in 1957, only a few people in the secret conclaves realized that an even greater turning point had been marked: the Soviet Union had launched Sputnik 1 by a rocket which had been developed as an Intercontinental Ballistic Missile. It is fortunate for mankind that this public demonstration of the advanced state of Soviet science and technology was made with a scientific payload in the nose of the rocket and not with an atomic bomb.

The history of those twelve years, from 1945 to 1957, is packed with the most remarkable examples of the scientific-military interface. It seems scarcely credible that the USA, which in April 1945 acquired von Braun, his technical experts, fourteen tons of technical documents about the V2, one hundred V2 rockets that were shipped to New Orleans and the manufacturing plant in the Harz mountains, should nevertheless yield precedence to the USSR in the production of the ICBM and the launching of an Earth satellite. The essential reasons lie in the single-minded determination exhibited by the USSR, coupled with a confusion and conflict of interests in the USA.

The Soviet Union acquired some of the Peenemünde team whom they set to work to improve the V2. After a successful test in 1947, this team was repatriated to Germany because the Soviet chief designer Sergei P. Korolev had by that time decided that the development of the V2 technology was not the way to create a Soviet long-range missile. Backed by the military planners, whose united aim was to produce a rocket capable of lifting their heavy primitive atomic bomb, Korolev and his team dealt so brilliantly with the problem that by April 1956 an intermediate range missile was tested. In August 1957 the ICBM was tested and it was this which led Khrushchev to announce that "a super long-distance intercontinental multistage ballistic missile was launched a few days ago. The tests of the rocket were successful. They fully confirmed the correctness of the calculations and the selected design. The missile flew at a very high unprecedented altitude. Covering a huge distance in a brief time, the missile landed in the target area. The results obtained show that it is possible to direct missiles into any part of the world."[4] It was this rocket —a missile developed to launch a heavy atomic bomb—which placed Sputnik 1 in orbit around the Earth on 4 October 1957.

Of course, a major requirement of the military planners in the USA during those years was also to produce an intercontinental missile. In the event, the United States intermediate range missiles, Jupiter and Thor, were a year behind the Soviet equivalent, and the Atlas ICBM lagged at least eighteen months behind the Soviet version which launched the Sputnik. There are complex, interlocking reasons for this situation. First, there was intense interservice rivalry between the United States army, navy and air force, each having its own missile development program and thereby eroding the limited funds which were made available in those years. Second, the military planners did not believe that it was necessary to invest in a rocket capable of carrying the heavy atomic bomb;

judging that the much lighter H-bomb would soon be available, they preferred to wait until a less expensive but more devastating ICBM could be produced. Finally, and perhaps the most serious, neither the Defense Department nor the scientists understood the rapid advance of Soviet science and technology in the years after 1945. General James M. Gavin has given a revealing account of the attitudes in America in this pre–Sputnik era.[5] He is critical of Secretary of Defense Charles E. Wilson whose comment, when told of the launching of the Sputnik, was that it was a "neat scientific trick."

These attitudes were exacerbated by a surprising reaction to pressures from the scientific community in October 1954 to launch a small satellite vehicle for scientific observations during the International Geophysical Year, which had been designated from July 1957 to December 1958 to cover the period of the maximum phase in the eleven-year sunspot cycle. On 29 July 1955, President Eisenhower announced that the United States would launch an Earth satellite during this IGY. One day later a similar announcement was made by the Soviet Union. The general reaction of excitement, coupled with a disbelief in the Soviet capability, bore no relation to the state of the projects in the two countries at that time. The Soviet scheme was based on the use of the rocket developed for military purposes. In the USA, distinguished scientists, led by von Braun, had recommended in the summer of 1954 that the American satellite should be placed in orbit by using an army Redstone missile together with a cluster of three solid-fuel rockets (a Jupiter C vehicle). This scheme, known as Project Orbiter, was rejected in September 1955 by an advisory group under Assistant Secretary of Defense Donald A. Quarles, in favor of a proposal that a group in the Naval Research Laboratory should develop an entirely new satellite launcher. The proposal of the US air force to use the Atlas ICBM then under development was also rejected. These decisions and the ruling by President Eisenhower that the development of the US satellite under the code name Project Vanguard should not interact with, nor divert any effort from, the

US ballistic weapon development led to the series of disasters which afflicted the project. The minimal goals set for Vanguard, the low priority and lack of financial provision, meant that the US effort to place a satellite in orbit had to be based on the technology of sounding rockets* instead of on the mainstream ballistic rocket techniques. The director of Vanguard was even refused permission to use the army testing facilities at Cape Canaveral on the grounds that this would interfere with the missile test program.

The failure of Vanguard in the shadow of the immense success of Sputnik ended the American dream that scientific space research could be a civilian enterprise. Von Braun again pressed his case that the army ballistic missile should be adapted to launch a satellite. Within three weeks of the Sputnik he received permission and money to do this, and at the end of January 1958, Explorer 1 was successfully placed in orbit.

THE LESSON OF SPUTNIK 1

For the world in which we live today there are critical lessons to be learned from this brief account of the circumstances surrounding the historic launchings of the first Earth satellites. President Eisenhower and his advisers correctly judged that the proposal to launch an Earth satellite was an entirely peaceful proposal arising from the decision of the International Council of Scientific Unions to designate an International Geophysical Year for simultaneous worldwide observations by the scientific community. Their understandable judgment was that the use of a military rocket to place such a satellite in orbit, with the constraints and secrecy classification which such a military association would imply, would be harmful to the peaceful intent of the satellite pro-

---

*That is, rockets designed to shoot small scientific payloads to altitudes above about 100 km for only a minute or so, and without the much greater rocket power required to give the necessary thrust so that the payload could achieve sufficient velocity to be injected into orbit around the Earth.

ject. They failed to recognize two vital factors: one, that the Soviet Union would exhibit no such scruples about the use of a military rocket, and furthermore, that they were more advanced than the United States in the development of a suitable rocket; two, that the priority and amount of money allocated to Vanguard ($11 million)* was quite inadequate for such a venture.

Although in the crisis created by the Sputnik, von Braun was given only $3.5 million to convert the army missile to a satellite launcher, immense sums had already been spent on its development. Billions of dollars were being spent simultaneously by the army, navy, and air force on rocket technology. The USAF spent $17 billion on the Atlas ICBM (this rocket was used in the first US attempts to send a probe to the Moon in 1958). The sums of money with which the US was prepared to back the Vanguard project in the vital period 1955–1957 were a thousand to ten thousand times too small. It is, of course, true that a limited form of space research could have developed on a purely civilian basis, but this would have been entirely different in character from that to which we have become accustomed. Sums of the order of $100 million in 1957 could produce a miniature Vanguard satellite—a payload consisting of a 6-inch sphere weighing 3.25 pounds was launched on 17 March 1958—but the billions of dollars necessary to develop the rockets which launched the Sputnik and Explorer satellites were available only within the context of defense requirements.

I have analyzed elsewhere[6] the extraordinary and dynamic influence of the Sputnik on American research and development organizations. In one year the R and D expenditure was doubled. It increased by three times in the decade before 1966. The actual expenditure on US space activities, both civil and military, was $59.9 million in 1955. Six years later it was $1,779 million—the civil expenditure on space increased by about fifteen times and the military by over two hundred

---

*The final cost of the whole Vanguard project was $110 million.

times in this period.[7] But even this was insufficient; the USSR forged ahead in the remarkable exploits of lunar probes and men in space. It was these extraordinary public demonstrations by the USSR of their newfound scientific and technological prowess, with all the accompanying military associations, which stimulated President Kennedy to make his famous speech to Congress of 25 May 1961: "We have examined where we are strong and where we are not, where we may succeed and where we may not. Now is the time to take longer strides—time for a great new American enterprise—time for this nation to take a clearly leading role in space achievement which in many ways may hold the key to our future on Earth." The US budget for civilian space activities rose to over $5 billion by the mid-1960s, and the brilliant feat of landing men on the Moon and returning them safely to Earth was accomplished at a cost of more than $25 billion.

THE DOCTRINE OF STRIFE AS JUSTICE

Two thousand years ago the Ionian philosopher Heraclitus revealed his contempt for mankind. He believed that "war is common to all and strife is justice, and that all things come into being and pass away through strife." Heraclitus would surely feel that the twentieth century exemplified his doctrine as no other era had done. The torment of World War II stimulated a massive technological development of discoveries about the nature of the atom made a few years previously during the peaceful inquiry into the deepest nature of the natural world. Within a few years the weapons were forged which destroyed Hiroshima and Nagasaki, and the ballistic rocket as a bomb carrier was developed. It is fortunate for the human race that the participants on one side or the other in that conflict did not discover how to make both the atomic bomb and the rocket. Now, only a little over three decades later, we exist in a fantasy world at peace; where material power is realigned so that East faces West, each grouping poised to strike with hydrogen bombs carried by rockets.

The attempt of the Eisenhower Administration in 1955 to divert space research into civilian channels by initiating the Vanguard project failed. Even without the Soviet confrontation we have seen that the containment of space research within a civilian budget could have led only to forms of scientific research in space of a minor character compared with those that exist today. A year after the Sputnik, when both the failure of Vanguard and the success of the military involvement were manifest, President Eisenhower made a second attempt to separate US space research from military associations. He accepted the recommendation of his Scientific Advisory Committee under James R. Killian that a strong civilian-oriented agency should be created to direct the scientific exploration of space. The National Aeronautics and Space Act became law on 29 July of that year. Within a few years NASA had absorbed eight thousand employees of the National Advisory Committee for Aeronautics, the Vanguard team and the Naval Research Laboratory rocket group, the 2,800 strong Jet Propulsion Laboratory of the California Institute of Technology, and von Braun's army group with a staff of 4,600. The independent space activities of the Armed Services—particularly those of the USAF, and the use of the Atlas ICBM for the lunar probes—were soon absorbed and superseded by this civilian agency.

The success of NASA is beyond question. As a civilian organization, pledged to international collaboration, NASA has had a major effect on space science and especially on the advance of astronomy. By his act of statesmanship, President Eisenhower made a significant separation of the defense and scientific interests in space. This successful development must not be allowed to obscure the fact that today, civilian space research represents a rather small proportion of the total space activities. Naturally, the security surrounding military activities, particularly in the USSR, creates difficulties in making any accurate assessment of the number and nature of the payloads sent into space. Nevertheless, it is impossible to screen the presence of a payload in orbit from modern radars, and with simple receiving equipment it is possible to study the nature of the orbit of a spacecraft and derive considerable information about the purpose of space missions from the received signals. Official US publications provide important information on the basis of such reports, and the following table shows a comparison between the US and USSR of the number of successful space launchings which were "primarily civil-oriented" or "presumptively military-oriented."

	USA			USSR		
	Total	Civil	Military	Total	Civil	Military
1957–1974[8]	621	304 (49%)	317 (51%)	789	265 (34%)	524 (66%)
1975[9]	28	19 (68%)	9 (32%)	89	27 (30%)	62 (70%)
1976[10]	26	15 (58%)	11 (42%)	99	24 (24%)	75 (76%)
1957–1976	675	338 (50%)	337 (50%)	977	316 (32%)	661 (68%)

According to Charles S. Sheldon II, who compiled the data for 1957–1976 [8, 9, 10] fifty of the USA launchings in 1957–1976 listed under "civil" were under the auspices of the Department of Defense. Thus to the end of 1976 about three-fifths of space launchings by the USA and USSR were either under the control of military authority or for military purposes. (This reservation is necessary because it is believed that all Soviet space launchings are under the control of the Soviet Strategic Rocket Forces.) The purposes of this extensive military activity are primarily reconnaissance, surveillance, navigation, electronic intelligence monitoring, and communications. Some considerable concern has been expressed, however, about the evidence that since 1976 the USSR has perfected a fractional orbital bombardment system, and has made significant progress toward active warfare in space by the successful maneuvering in space of two satellites leading to interception and destruction of one of them.

There has been little official comment on the detailed nature or value of defense activities in space, apart from a speech made by President Johnson at Nashville, Tennessee, on 15 March 1967, in which he claimed that the dividends for the US of military space operations were equal to ten times everything which had been spent on them. This statement implied that there had been savings in the US defense budget of $400 billion over a ten-year period because more precise information than could otherwise have been obtained about military and military-related activities all over the world was received through space operations.

In 1973 there was a classic revelation of the interlocking of scientific and military interests in space. The US Department of Defense launched a number of spacecraft under the code name Vela. Four spacecraft were equally spaced in circular orbits, at distances of 100,000 km from Earth, designed to detect any nuclear explosions made by a foreign power. Between 1969 and 1972, these spacecraft detected sixteen bursts of γ-rays from 0.1 to 30 seconds in duration, and when it was realized that these bursts did not come from the Earth but from some unknown source in the Universe, permission was given to publish the information.[11] A civilian-operated satellite IMP-6 was simultaneously in orbit with the precise purpose of detecting γ-rays from space. It did not discover this remarkable, and so far inexplicable, astronomical phenomenon, although in retrospect it was found that the bursts had in fact been detected by the IMP satellite.

The evolution of the ballistic rocket has had a complex effect on human affairs. On the one hand man has been provided with spectacular possibilities for probing the mysteries of the cosmos and for gaining an understanding of his place in the Universe. On the other hand the civilized world is now, for the first time in human history, equipped with the means of total destruction. And the dilemma for man is deepened by the appreciation that the functions of the equipment are effectively interchangeable. Since 1957, more than half of all the payloads sent into space have had a military connotation. The rockets which placed the payloads into space, both civil and military, have been intercontinental ballistic missiles or closely related to such weapons.

An important analysis of the types of rocket which have

been used to place the 7.7 million kilograms of payload in orbit, to the end of 1975, has been published by the Committee on Aeronautical Space Sciences of the United States Senate.[12] The conclusion is that the original Soviet ICBM (the S.S.-6, Sapwood missile) has been the most frequently used launch vehicle in the world, followed by the US intermediate range missile Thor. A limited number of rockets have been designed especially for launching primarily nonmilitary payloads, either because of their small thrust (the US Scout) or because the requirements are beyond military needs (e.g. the USA Saturn-manned Moon flights). This analysis gives the following totals for the two categories.

Total Launches 1957–1975

	Total launches	Number by military missiles	Number by nonmilitary missiles
USSR	878	837 (95%)	41 (5%)
USA	655*	562 (86%)	93 (14%)

The confrontation of man's creative and destructive activities is nowhere more clearly evident than in this dominance of military interest in space—to the end of 1975 the combined USA and USSR payloads in orbit were 60 percent military, and the rockets which launched the civilian and military payloads were 90 percent ballistic missiles.

There is little hope of change in this situation in the foreseeable future. The major emphasis on new developments in the United States is on the reusable space shuttle, which will certainly dominate space science for the remainder of the century, and, if the plans materialize for the large space telescope to be launched by using the shuttle, it will be the preeminent observational instrument for astronomers. The following comment of the Chief of the Science Policy Research Division of the Library of Congress, a foremost international authority on the Soviet and American space activities,[13] is

*This number, taken from Table 1.7, page 40 in vol. 1, chapter 1 of *Soviet Space Programs, 1971–75*, differs by six from the data quoted in the previous table (page 140), taken from C. S. Sheldon.

therefore significant: "Until the US shuttle, intended for use by both NASA and the Air Force, becomes operational in about five years, we can expect the Soviet position of leadership to be further consolidated. . . . Should the Soviet Union be successful in developing a reusable space shuttle in close to the same time frame as the United States, the hope that the US shuttle will necessarily change the balance in the favor of the United States might not be realized. Worse, if there were no American shuttle and a Soviet reusable system appeared, it is possible that a whole new set of rules on the use of space might be written by the *de facto* situation that would develop. The space equivalent of freedom of the seas might be lost in the face of overwhelming Soviet dominance in volume and variety of their operations. In the meantime, it is not surprising that US military planners are paying increasing attention to the development of defensive techniques to increase the survivability of US spacecraft. The nation relies on space to provide so many vital services of communication, navigation, weather reporting, early warning, and arms-agreement policing that it would be seriously damaged if these functions could be terminated or greatly reduced by the act of another power."

The confrontation in space is a most perplexing and perilous aspect of the contemporary world. The scientific and peaceful applications of space research and technology provide the human race with immense opportunities for intellectual and material advances. At the same time, several thousand of the same launching rockets, equipped with atomic warheads of explosive power measured in megatons are poised for a destructive act which could lead to human disaster on an unimaginable scale, exceeding by far the greatest of the world's natural catastrophes, the Shen-Shu earthquake of 1556 which wiped out a million people in a few seconds. We are like King Lear in his madness: "I will do such things— what they are, yet I know not; but they shall be the terrors of the Earth."

IX

Human Purpose and the Progress of Civilization

It is now more than half a century since A. N. Whitehead expressed his opinion that "when we consider what religion is for mankind, and what science is, it is no exaggeration to say that the future course of history depends upon the decision of this generation as to the relations between them."[1] His point was that religion and science were the two strongest general forces influencing man, and that the forces of religious intuition and the force of the impulse toward accurate observation and logical deduction seemed at that time, in 1925, to be set one against the other.

Whitehead spoke at a time of hope and determination that a cataclysm like World War I should never again descend on humanity. Even so it was a time of divisive opinions about the future. Modern science had demonstrated its great power in warfare and because of this there was disquiet and widespread discussion about the future of civilization in these new circumstances, with science and technology advancing more rapidly than ever before. Hope and determination were already in conflict with the disruptive forces of society. On the one hand, in 1926, Germany was admitted to the League of Nations; on the other, the American Senate had opposed the League and had never ratified the Treaty of Versailles in which, at Woodrow Wilson's insistence, the League had been incorporated. The tireless work of Wilson, who inspired the formation of the League, was of no avail because the country of which he was President never became a member.

Such divisions were not new in the history of mankind. Almost the whole of recorded history is centered around the use of available material power in conflict—and frequently in the name of religion or in support of religious faith against the evils of decadence. The romantic vision of past ages should not obscure the legacy of historical evidence which demonstrates that, while the material power of a society has occasionally been deployed for the benefit of that society, more often it has been wielded by the rulers for expansionist and malignant purposes. Vast empires have been created by force in the name of freedom and religion, and eventually they have been destroyed by force or decadence. The present state of mankind has been determined by the unstable, complex interaction of the intellectual power of a rather small number of individuals, coeval with the group instincts of greed.

In the sixth century B.C., for example, Cyrus the Great of Persia inherited a great empire from his father. When he died fighting ten years later his empire extended from the Caucasus to the Indian Ocean, and from Ionia to Central Asia. By 522 B.C. his son had completed the conquest of Egypt, but the huge empire was unstable and Cambyses, the son of Cyrus, was killed in rebellion. Darius came to power, and resuming the expansionist policy of Cyrus he crossed the Indus and invaded the Punjab. Although by the turn of that century he ruled over one of the greatest empires in history, he desired to add Europe to his dominions. The great army of Darius was defeated at Marathon in 490 B.C. and the fleet of his successor, Xerxes I, at Salamis ten years later. Soon, the entire Persian Empire was conquered by the Macedonians. Yet these were the ages as well of the great Ionian philosophers, of the revolutionary cosmologies of Anaximander, Anaximenes and Anaxagoras. Pythagoras of Samos, who discovered the famous proposition 47 of Euclid, and that the concordant intervals in a musical scale are produced by simple numerical ratios, lived at this time (from about 570 to 490 B.C.), founding a school which existed for nearly a millenium. While Darius was preparing to conquer Greece, Pythagoras was preaching the metempsychosis of souls and the purification of the soul through knowledge of music, arithmetic and geometry.

At every point of history we find these extraordinary contrasts: the coexistence of intellectual excellence with the destructive forces of a decadent materialism. The past has never contained excellence untarnished with evil and it has always been a failure of courage to contrast the excellence of the past with the misfortunes of the present. We praise Pythagoras but do not associate his age with that of Cyrus and Darius. For those who lived in the late Middle Ages, the golden age was already in the past: the Greek and Roman Empires had decayed; Aristotle and Virgil were long since dead. They were engulfed by bitter wars and religious conflicts. They could not know that Copernicus, born in 1473, would start an intellectual revolution such as the world had never before witnessed, and that in their age the Aristotelian doctrine would be undermined. Neither could they know that new technology applied to ships, facilitating the voyages of Columbus, would end the economic and intellectual isolation of Europe. The sacking of Mainz in 1462 seemed a more important event than the presence among the citizens of Johann Gutenberg, the inventor of movable printing type.

The critical point is that man has survived, progressed and prospered—that is, considered over the time span of centuries and millenia. It is, and always has been, impossible to assess the importance of any age to the progress of civilization within the context of current events. We have no idea how the twentieth century will appear in the historical context. We do not yet know whether the world wars, the atomic bomb, the rockets, and the multitudinous paraphernalia of modern war will appear as significant as Einstein or the Moon journey of Armstrong and Aldrin. Nevertheless, it is not unreasonable to speculate on the answer to the question raised by Whitehead a half century ago; after all, he believed then that the decision would rest with the generation whose work is now almost complete.

When Whitehead spoke about the force of religion he did not mean the force of any narrow dogma. He spoke of religion in the broadest sense: "Religion is the reaction of human nature to its search for God." When he spoke of the opposing force arising from the impulse to accurate observation, he could not have foreseen that the generation to which he referred would achieve new and devastating powers of destruction, nor that the impulse to logical deduction would simultaneously raise doubts about the limits and validity of such procedures and conclusions. Today it may appear that the middle decades of the twentieth century have created entirely new circumstances for mankind, that materialism, material power, national and personal avarice and greed may have triumphed over the opposing forces of religion, and the intellectual purposes of the civilized world. Is it yet possible to judge whether the pessimism is generated, as in every past age, because of the restriction to a localized view, or will the twentieth century be seen in history as the beginning of an irreversible decadence?

There are grounds for anxiety. Although mankind has survived the endless armed conflicts of the past, those of the twentieth century have attained a more global scale than in any previous age. The decisive battles of the past have been localized: from Marathon and Salamis to Quatrebas and Waterloo, we see vast territories and issues at stake, resolved in concentrated mortal conflicts. Furthermore, in the analysis of these conflicts it can be seen that elements of human courage and intelligence have often determined the victory. The immense Persian army at Marathon was slaughtered by the small Athenian army without even the help of the Spartans, who considered their religious festival to be of more significance. When the small fleet of the Greeks was hopelessly outnumbered by the thousand ships of the Persian fleet at

Salamis, and when the Greek commanders were in dispute, it was Aristides, banished from Athens seven years earlier, who came from Aegina bringing news of the disposition of the Persian fleet to his rival Themistocles: "Let us still be rivals but let our strife be which can best save our country." The ultimate defeat of Napoleon more than two millennia after Marathon occurred in a similarly localized conflict. On the field of Waterloo, Wellington engaged Napoleon's army at 11:30 on the morning of 18 June 1815. Nine hours later Napoleon was in flight.

The human slaughter of these battles—more than fifty thousand dead—bears no comparison with the casualties of twentieth-century conflicts. Ten million men were killed and thirty million wounded in the war of 1914–18. In the second global conflict, a quarter of a century later, even these figures were dwarfed. The Axis powers and the Allies together suffered 24.5 million killed in battle. And now war against civilians was part of the conflict; the civilian toll in World War II is estimated at thirty million dead.*

There can be no question that the forces of destruction available in the twentieth century have created a sharp discontinuity in human conflict. For thousands of years the advance of science and technology altered the tactics of warfare, but not the formal strategy. Now the aircraft, the rocket and the nuclear bomb have changed the dimensions to such an extent that in the event of future conflict the question is not which nation or society will be victorious, but whether human civilization can survive. In our pessimism we may well feel with Heraclitus that "all things come into being and pass away through strife." It is as though the nineteenth-century philosophy of Nietzsche, in which the man of the future was to be modelled "by means of discipline and also by means of the annihilation of millions," has come to fruition.

In a curious sense modern society appears to exhibit the

---

*This estimate includes the victims of the Chinese-Japanese war and the Jews exterminated by the Nazis.

purpose which governed the thought of Nietzsche. He believed that the strength of the human will is to be measured by "the amount of pain and torture it can endure and know how to turn to its own advantage." He did not deplore, as we do, the evil in society but entertained the hope "that life may one day become more evil and more full of suffering than it has ever been." Nietzsche predicted the era of great wars which came to pass soon after his death. But his ethics, based on the belief that the victors in war are biologically superior to the vanquished, belong to Sparta and not to Hiroshima and Nagasaki. It may not have been harmful in the mid-nineteenth century for Nietzsche to propagate the ethic which made the Spartans great, at least as seen through the eyes of Herodotus or Plutarch; but manifestly applied to our times it would be catastrophic. The total sacrifice of Thermopylae is no guide to the achievement of greatness in the modern world.

The pessimism of the modern age does not only arise through fear of extinction in war. There is, simultaneously, the global problem of survival in peace. The tortuous difficulty of the four billion members of the human race is exemplified by the statistic that more than one half of this number are either illiterate or diseased. It is an insufferable burden scarcely alleviated by the knowledge that one-tenth of the world's wealth is spent on arms, four times more than on education or health. Perhaps the Hegelian doctrine that in the only real time-process we proceed from the less to the more perfect and the consequential Marxian dialectic leading to the belief in the inevitability of progress have served to undermine the ethic of compassion. Today, with starvation rampant in the Third World, we face the dilemma that not only the Great Powers but also the governments of the Third World feel the need for weapons rather than tractors.* Often it appears that our contemporary civilization has become dominated by personal and national ambitions, excluding the

* *The London Times*, 10 September 1976, made the comment that the cost of a single tank could pay for a hundred food-producing tractors.

deeper ambitions for the understanding of human purpose.

The influence of scientific investigation and discovery on these problems during the last half century has been peculiar and the ultimate effects are not yet resolved. An immense optimism was engendered in the West by the brilliantly successful use of science in World War II. Science and scientists emerged from the conflict with a status in society entirely different from that which they had enjoyed in the prewar phase. Not only in the Soviet Union, but throughout the Western World, science and its technological applications appeared to be all powerful. For a time, there seemed to be a magic wand which, when used in the service of man, would solve all problems. Society was persuaded to give great material support to scientific activities in the belief that the discoveries would, inevitably, in some way, be of practical benefit to humanity. Already we see that this is not the case. The more science has progressed, so the divisions between the good and evil in its applications have vanished. The belief in automatic material progress by means of scientific discovery and application has become a tragic myth of our age.

The internal tragedy for science resides in the circumstance that public faith in this activity has changed its nature—the great financial requirements have eroded the possibilities for private sponsorship, and almost everywhere the conduct of research has become an important affair of state. Although there are still localized centers where science remains a neutral activity, centers whose survival and extension are a major necessity for the integrity of science and the progress of civilization, it is clear that the major part of scientific activity, and the whole of that activity demanding great resources, is interlocked within the framework of the society in which it is pursued. In October 1975, during the great two-hundred-fiftieth anniversary celebrations of the Soviet Academy of Sciences in Moscow, Leonid Brezhnev, the General Secretary of the Central Committee awarded the Order of Lenin to the Academy. In his speech[2] he said, "A Soviet scientist—if he is truly a Soviet scientist—proceeds in his entire research work from the scientific ideology of Marxism-Leninism, is an active fighter for the cause of communism, against all forces of reaction and obscurantism. Our scientists subordinate all their practical work to the task of implementing the noble idea of communism." In his famous speech to Congress on 25 May 1961, President Kennedy, in committing the nation to the greatest, most hazardous and expensive scientific undertaking in history, said, "We set sail on this new sea because there is new knowledge to be gained, and new rights to be won— and they must be won and used for the progress of all people. For space science, like nuclear science and all technology, has no conscience of its own. Whether it will become a force for good or ill depends on man." The speeches of Brezhnev and Kennedy surely epitomize the contrasting hopes and dangers and the unresolved function of science in the modern world.

Although there are grounds for deep anxiety, there are reasons for hope. The calamities of the twentieth century have parallels in history. The golden ages of the Roman Empire embraced periods of terrible decadence. The reign of Marcus Aurelius, in the second century A.D., was, like our own age, beset with natural and man-made disasters—pestilence, earthquake, interminable wars and anarchy. Men were fighting with beasts, the economy was in ruin and the Christians were persecuted. The *Meditations* reveal the burdens and weariness of Marcus Aurelius. Yet Gibbon considers the period to have been one when the human race was most happy and prosperous, and indeed, the Empire continued to flourish and more than two centuries elapsed after the death of Aurelius before Rome was sacked by the Goths in A.D. 410.

For the Western World the outlook for civilization seemed as bleak in the fifth century as it has ever done in the twentieth. Saint Jerome wrote in A.D. 413: "The world sinks into ruin: yes! but shameful to say our sins still live and flourish. . . . Churches once held sacred are now but heaps of dust and ashes: and yet we have our minds set on the desire for gain." It is a description which a modern saint might well apply to the world of the twentieth century.

The fifth century may have been full of disaster and deep pessimism, but it also contained the hope for the future: when Saint Jerome was writing his letters of dismay, Saint Augustine was simultaneously at work on *The City of God,* a book of immense importance for the survival of Christianity in the following centuries. His comprehensive scheme of Christianity provided the inspiration and hope for the future, and substantiated the doctrine that the Church was separate from the State, which must be submissive to the Church in all religious matters. Not even the Communist states of the twentieth century have succeeded in extinguishing the legacy bequeathed to the world by Saint Augustine. He epitomized the hope that must exist even in the most decadent periods of civilization if the human race is to survive and prosper. It is the hope, disconnected from the localized pessimism, that enables the contemporary evils to be endured.

The anxieties of the twentieth century have been largely centered on the simultaneous trust in and fear of science and technology. For a long time the trust was found to be justified. In those regions of the world where science and technology were applied to the betterment of the human race, great progress was made. The standards of health and comfort, those aspects of life affecting the physical well-being of the individual, have been immeasurably enhanced over those enjoyed by our forefathers. But after World War II, many problems arose. These have been of two kinds. First, it was found that science and technology could not be used in peace as in war. Although they were powerful and vital activities, the human society to which they were applied often had complex and unexpected reactions which defeated the purpose of the technological application. As distinct from a society at war, a society at peace was found to be diverse in its motives and purposes, and the injection of science and technology often accentuated the divisions. In extreme cases the technological injection designed to secure harmony and prosperity led to immediate disruption and economic difficulties. The second type of problem concerns the balances of prosperity on the global scale. For the poor and underdeveloped countries, the injection of science and technology seemed to be the elegant and humanitarian path toward easing the problems of poverty and starvation, but again, there has been failure because the human and political problems have proved greater than the scientific ones. It is extraordinary that the efforts to redress the balance between the rich and the poor nations through modern science and technology have not merely failed to do so, but have worsened the situation substantially. For example, the average income per capita of the Western industrialized countries compared with that for India was 10 to 1 in 1950. Twenty years later it was 40 to 1. The application of science to the problems of the world population have exacerbated the discrepancies, increasing, where they were designed to decrease, differences of wealth, power, of physical comfort and health. The sources of discord and the magnitude of the world's moral problems have been enhanced to the extent that the disparities now generate political tensions of explosive proportions.

These failures have increased the gravity of the national and global problems. On the other hand, the fact that failures are now recognized and demonstrated provides hope for the future, a hope which springs from the recognition that only those aspects of modern technology which can be integrated with the complex needs of the local society will be successful. In the utopia of Plato's *Republic,* justice consisted in every man doing his own job. The translation to the underdeveloped peoples of the modern world is simply that it is useless to build a computerized canning factory until individuals are provided with the means and purpose to till the ground and fish the seas.

The fear of modern science and technology has both a physical and spiritual basis. The physical basis of fear is of two types. First, there is the anxiety about destruction through armed conflict, which has been discussed. The grounds for hope in this case are peculiar: they reside in the cataclysmic nature of the weapons of modern war, in the awareness that no nation can hope to attain a classical victory in a modern global conflict. The dicta of preemptive strike, of first counterstrike or second strike are the dicta of human suicide. The hope for the future lies in the fact that a balance of power has been established through an era when there existed a major imbalance. Thus, after World War II, until August 1949, the United States possessed the atomic bomb and the Soviet Union did not; similarly, from 1952, the United States possessed the hydrogen bomb, while the Soviet H-bomb was not tested until August 1953. During these years there were groups in the United States who were in favor of a preventive war against the Soviet Union while the Americans had a stockpile of the bombs and the Russians had not. Nations can no longer calculate as Locke proposed in the second of his *Treatises on Government* that in defense one should debate the amount of injury which may justly be inflicted upon the aggressor.

The second type of physical fear arising from modern science and technology is that the natural environment of the Earth may be irreparably damaged, not through war but in peaceful activities. The complex interrelations between organic life, water, soil and the atmosphere may be unbalanced so that the ecosystems collapse irretrievably, for example by the cessation of photosynthesis, or of the fundamental chemical-biological cycle.

The history of the human race gives grounds for disquiet. Twelve thousand years ago the few million people living on Earth appear to have hunted many animal species to extinction. Two hundred years ago the billion inhabitants of Earth began to devastate the forests, erode the soil, and pollute the oceans. Today, four billion people are using the Earth's natural resources at an alarming rate and pouring vast amounts of pollutants into the atmosphere and oceans. The disquiet is increased by the knowledge that the Earth's human population reached only five hundred million by A.D. 1650 after six hundred millennia of evolution. In the next three hundred years it increased six times. The future projections are that it will double again by the early twenty-first century and reach about fourteen billion by A.D. 2014—unless there are great human disasters or enforced population controls.

Already the United States and the Soviet Union inject into the atmosphere every year 274 million tons of dust and other gaseous pollutants.[3] At least on the local scale there have been significant changes in climate, and there are authoritative estimates that if energy consumption continues at the present rate, the injection of carbon dioxide and other matter into the atmosphere may cause global temperature increases over the next fifty to one hundred years. The contemporary awareness of the problem is a source of hope, and the evidence for the stability of the atmosphere over ten thousand years engenders confidence that the terrestrial system is robust enough to counteract any perturbations likely to be introduced by man.[4]

The spiritual basis of the fear of science has a deeper justification than the anxieties about its physical effects. The ancient struggle to understand the world, and the question of whether such an understanding can be reached without the notion of purpose, has appeared to many to have been resolved by twentieth-century science. The problem for our age has been dangerously simplified—as indeed Whitehead so correctly envisaged—as a conflict between two opposing forces.

Modern science, and the almost unimaginable successes of the scientific method in the effort of understanding, have focused, as never before, the conflict of Socrates, and of Plato and Aristotle, with the atomists of the fifth century B.C. The great exponents of atomism at that time, Democritus and Leucippus, maintained that everything happens in accordance with natural laws. According to Leucippus, "Naught happens for nothing, but everything from a ground and of necessity." For the atomists the totality of existence was explicable without the need to introduce the notion of purpose. Their universe was mechanistic, and all questions about existence could be asked and answered in mechanistic terms. But it was the opposing view, especially of Aristotle, that purpose was a part of scientific procedure and governed the development of the Universe, which prevailed. Two thousand years later when the theory of Copernicus was accepted the philosophy of atomism was revived, mainly through the recovery of the work of the followers of Democritus and Leucippus, Epicurus and Lucretius, particularly the *De Rerum Natura* of Lucretius. The corpuscularism of seventeenth-century science achieved an uneasy but nevertheless fairly stable compromise, exemplified in the determinism of Newtonian theory coupled with his belief that "God in the Beginning, form'd Matter. . . ." For another two hundred years the contrasting beliefs remained largely disconnected. While Laplace was describing the mechanistic universe in his *Mécanique céleste*, the Puritans in New England were seeking understanding by running from one hour-long sermon to another.

This uneasy compromise between two conflicting attitudes toward existence and man's purpose was not destined to survive the nineteenth century. The British Association for the Advancement of Science had been founded in 1831 "to obtain a greater degree of national attention to the objects of science . . . and to promote the intercourse of the cultivators of science with one another, and with foreign philosophers." These objects were achieved in a dramatic and unexpected manner when the Association met in Oxford in 1860. Charles Darwin's *Origin of Species* had been published in 1859 and the appropriate section of the Association arranged a discussion of the Darwinian theory. The antagonisms were such that the discussions were continued two days later before a very crowded meeting during which Bishop Wilberforce of Oxford turned to T. H. Huxley and inquired whether it was through his grandfather or grandmother that he claimed descent from a venerable ape. Huxley replied, "If I am asked whether I would choose to be descended from the poor animal of low intelligence and stooping gait, who grins and chatters as we pass, or from a man, endowed with great ability, and a splendid position, who should use these gifts to discredit and crush humble seekers after truth, I hesitate what answer to make. . . ."

This violent, public confrontation between science and orthodoxy was an occasion of great significance in the erosion of the transcendental view of the world "sky-woven and worthy of a God" which had been derived through millennia of religious thought and activity. The victory of the supporters of Darwinism over the poorly formulated attack from a prominent Anglican bishop enormously increased the prestige of science as the means through which total knowledge should be sought. For the next century science seemed all powerful. The technical developments which made possible the penetration into the depths of the Universe, the investigations into the structure of the atom and the comprehension of the observed phenomena within the framework of new and revolutionary theories led to the belief that everything which could be observed would have a scientific explanation. The battle lines between science and orthodoxy became rigidly established and when Whitehead said that the two forces seemed to be set one against the other, he may not have realized the extent of the tactical victory already achieved by the forces of science. For that generation, science coupled with technology became the God through which man was seeking the road to economic and intellectual salvation.

Now, as we enter the last decades of the century, we see the fallacies in this long battle. Observable phenomena *are* subject to investigation by the tools and methods of science. The immediate answer may be only partially correct but there is no reason to question that, given time and technique, knowledge in the language of science can be achieved. At last in this process of scientific probing we find that the investigations have revealed the existence of phenomena which are neither observable nor deducible from those aspects which can be observed.

Here we have treated this in terms of one of the great intellectual problems always confronting man. The evidence is clear that throughout our recorded history one purpose of man has been to seek solutions to the problem of the beginning. In the first half of the twentieth century the problem apparently became a definitive scientific one. We see now that the observational evidence most widely accepted is that the Universe is expanding, and that there *is* a problem of a beginning—a creation—of matter and of man's existence in this particular universe. We have no scientific description of the beginning. It may be that, eventually, a more detailed understanding of the observations will place a different interpretation on the redshift, and thereby we can evade the difficulties of the expansion and of the concentrated beginning. In that case we face the problem of creation at some point in an infinite past time, and again this is not observable nor describable in scientific language. It is difficult to see how under these circumstances one can advance on the opinion of Saint Augustine, who in answer to the question of why the world was not created sooner replied that there was no "sooner" because time was created when the world was created. This, of course, is the essence of general relativity in which neither time nor space existed before the singularity. It seems that modern science has created a real dilemma in the search for an understanding of the ultimate past of the Universe. The conclusion is that reached by Wittgenstein in his attempt to discover a logical system for understanding the conclusions of scientific and mathematical analysis: "What we cannot speak about, we must pass over in silence."[5]

The pursuit of understanding is an essential occupation and purpose of modern society, as it has been throughout all the thriving societies of history. The prominent fear of the mid-twentieth century that science was all powerful and would achieve the ultimate finality in these pursuits is now seen to be without foundation. The immense intrinsic values of science and of the scientific method to the modern community are well established, and will be enhanced by the recognition that they are absolute neither in practical nor intellectual pursuits. The rediscovery of the validity of non-scientific modes of thought and investigation will establish once more the faith of the people in science itself, to the detriment of the pseudosciences which tend to gain monopoly in periods of decadence. In a recent address to the American Academy of Arts and Sciences,[6] the distinguished physicist V. F. Weisskopf quoted the words of Marcus Fierz, the Swiss physicist-philosopher: "The scientific insights of our age shed such glaring light on certain aspects of the experience that they leave the rest in even greater darkness." This is the answer to the grave issues raised by Whitehead fifty years ago. We can apply the spectroscope to gain an understanding of the sunset; we can send the space probe to Venus, but we may never apprehend the ethos of the evening star.

Notes

Chapter I. The Dilemma of Mankind

1. Thomas Carlyle, *Sartor Resartus*, written in 1831, published in 1838. The quotation is from Book I, 10, *Pure Reason* (vol. IV of the standard edition of Carlyle's works [London: Chapman and Hall, 1838].

Chapter II. The Evolution of Planetary Systems

1. From Melancthon's *Initia Doctrinae Physicae*. Translation by A. D. White in his *A History of the Warfare of Science with Theology in Christendom* (New York: Appleton-Century-Crofts, 1896), p. 126.
2. Ibid., pp. 126–127.
3. T. S. Kuhn, *The Copernican Revolution* (Cambridge: Harvard, 1957), Chapter 6.
4. A contemporary investigation of Galileo's conflict with the Church may be found in *The Crime of Galileo* by G. De Santillana (Chicago: University of Chicago Press, 1955).
5. Sir Harold Jeffreys in Proc. Roy. Soc. 214 (1952): 281.

Chapter III. Investigation of the Solar System

1. See, for example, the account of this period by A. Koestler, *The Sleepwalkers* (New York: Macmillan, 1959), Chapter 8.
2. D. H. Menzel and F. L. Whipple, Publn. Astr. Soc. Pacific, 67, 161, 1955.
3. C. H. Mayer, T. P. McCullough and R. M. Sloanaker, Astrophys. Jr., 127, 1, 1958.
4. Quoted in *Sky and Telescope* 52 (December 1976): 406.

Chapter IV. The Doctrine of Many Possible Worlds

1. *Nature* 184 (1959): 844. At that time Cocconi was on leave at CERN (European Organization for Nuclear Research) in Geneva, and Morrison at the Imperial College of Science and Technology in London. Some months earlier—on 29 June—I had received a letter from Dr. Cocconi on the same topic, urging me to use the radio telescope at Jodrell Bank in the search for extraterrestrial signals. The text of this letter was published as an Appendix to my book *The Exploration of Outer Space* (Oxford University Press, 1962), p. 82.

2. The book *Earth* by F. Press and R. Siever (Berkshire: W. H. Freeman, 1974) provides a clear account of modern views on the formation of the Earth and its atmosphere. More details of the probable chemical processes involved may be found in Chapter 11 of *Evolution of Sedimentary Rocks* by R. M. Garrels and F. T. Mackenzie (New York: W. W. Norton, 1971).

3. See, for example, the calculations in the paper by S. I. Rasool on evolution of atmospheres on the Earth and planets in Chapter 12 of *Physics of the Solar System*, NASA Document SP-300, Washington, D.C., 1972.

4. In a communication to the Russian Botanical Society in Moscow in 1922. A complete account of Oparin's work may be found in the English translation of his book *The Origin of Life on the Earth*, 3rd ed. (Edinburgh: Oliver and Boyd, 1957). The first and second editions of this work appeared in Moscow in 1936 and 1941.

5. See, for example, his book *The Inequality of Man* (London, 1932).

6. J. D. Watson and F. H. C. Crick, "A Structure for Deoxy-Ribose Nucleic Acid in *Nature* 71 (1953): page 737.

7. R. S. Edgar and W. B. Wood, "Morphogenesis of bacteriophage T4 in extracts of mutant infected cells." Proceedings of the National Academy of Sciences, 55, (1966): 498.

8. An authoritative account of these problems as they appeared in 1970 may be found in the book *Chance and Necessity* (New York: Alfred A. Knopf, 1971) by Jacques Monod, the director of the Pasteur Institute in Paris who was awarded the Nobel Prize in 1965 for his work in this field.

9. *Opticks*, 2nd ed. (London, 1717), page 377. In this passage Newton was, of course, referring to the problem of finding an explanation for the force of gravity.

Chapter V. The Nature of the Universe

1. An account of Herschel's telescopes derived from an investigation of the archives of the Royal Astronomical Society has been given by J. A. Bennett, "On the power of penetrating into space: The telescopes of William Herschel," *Journal for the History of Astronomy* 7 (1976): 75.

2. The typescript of Shapley's address is in the Shapley archives at Harvard University. It is reprinted in *Journal for the History of Astronomy* 7 (1976): 175. The figure of 60,000 light years given by Shapley as the Sun's distance from the galactic center was later found to be too great. The figure accepted today is 33,000 light years.

3. Mon. Not. Roy. Astr. Soc., vol. 5, no. 4 (14 February 1840), page 32.

4. Mon. Not. Roy. Astr. Soc., vol. 5, no. 12 (12 February 1841), page 89.

5. The lecture, given on 22 May 1908, was reprinted in *Royal Institution Library of Science: Astronomy*, vol. 2 (Netherlands: Elsevier, 1970), p. 78.

6. *Principia rerum naturalium sive novorum tentaminium phenomena mundi elementaris philosophice explicandi* (Dresden and Leipzig, 1734).

7. J. Michell, Phil. Trans. Roy. Soc. (1767) page 234, and (1784) page 35.

8. Ibid., (1785), vol. 75, page 213.

9. Ibid., (1791), vol. 81, page 71.

10. *Westminster Review*, vol. XIV, 1858.

11. Reprinted in *Royal Institution Library of Science: Astronomy*, vol. 1, (Netherlands: Elsevier, 1970), page 42.

12. Agnes M. Clerke, *The System of the Stars* (London: Black, 1890). Subsequent editions of this work maintained this view for the next fifteen years.

13. This analysis was published in 1975. See R. J. Dodd, D. H. Morgan, K. Nandy, V. C. Reddish, and H. Seddon, Mon. Not. Roy. Astr. Soc. 171 (1975): 329.

14. See B. Lovell, *Nature* 264 (1976): 506.

15. Willem de Sitter (1872–1934) was appointed professor of theoretical astronomy in the University of Leiden in 1908, and director of Leiden Observatory in 1919.

16. E. Hubble, Proc. Nat. Acad. Sci. Amer. 15 (1929): 168.

17. E. Hubble and M. L. Humason, Astrophys. J. 74 (1931): 43.

18. Published as *The Realm of the Nebulae* (London: Oxford University Press, 1936).

19. A. Sandage, Q. Jr. R. A. S. 13 (1972): 282.

Chapter VI. The Origin of the Universe

1. These results were published simultaneously by R. H. Dicke, P. J. E. Peebles, P. G. Roll, and D. T. Wilkinson, Astrophys. J. 142 (1965): 414, and A. A. Penzias and R. W. Wilson, Astrophys. J. 142 (1965): 419.

Chapter VII. Space: Creation and Comprehension

1. See, for example, S. G. F. Brandon, Creation Legends of the Ancient Near East (London: 1963), Chapter 2.
2. "Theogony" (eighth century B.C.) in a hymn of praise to Zeus.
3. See, for example, L. N. Cooper in The Physicist's Conception of Nature, ed. J. Mehra (Boston: Reidel, 1973), p. 668.
4. S. W. Hawking in Confrontation of Cosmological Theories with Observational Data, IAU Symposium No. 63, held in Poland in 1973 to commemorate the five-hundredth anniversary of the birth of Copernicus. Published by Reidel, 1974. Page 283.
5. R. H. Dicke, Nature 192 (1961): 440.
6. Summaries of recent work by many of the theorists now engaged in these researches may be found in two symposia volumes: Seventh Texas Symposium on Relativistic Astrophysics, Annals of the New York Academy of Sciences, vol. 262, 1975; I.A.U. Symposium No. 633, Confrontation of Cosmological Theories with Observational Data, pub. by Reidel, 1974.
7. Confrontation of Cosmological Theories with Observational Data, op. cit., page ix.

Chapter VIII. The Confrontation Between Man's Creative and Destructive Activities

1. Published by Space Publications, Inc., Washington, D.C.
2. Quoted by General Sir James Marshall-Cornwall in Early Rockets, Proc. Roy. Artillery Historical Soc. 3, (1972): 48.
3. Wernher von Braun and Frederick I. Ordway III, The History of Rocketry and Space Travel (New York: Thomas Y. Crowell, 1975).
4. The New York Times, 27 August 1957.
5. James M. Gavin, War and Peace in the Space Age (New York: Harper & Row, 1958).
6. Sir Bernard Lovell, The Origins and International Economics of Space Exploration, (Edinburgh, Scotland: Edinburgh University Press, 1973; New York: J. Wiley, 1973). Sir Bernard Lovell, "The Effects of Defense Science on the Advance of Astronomy," in Journal for the History of Astronomy, vol. 8, page 151, 1977.

7. These remarkable figures were published by the United States Bureau of Budget, 30 October 1961.

8. See Table 6, page 45, of *United States and Soviet Progress in Space: Summary Data through 1974 and a Forward Look.* Library of Congress, Congressional Research Service Paper, Washington D.C., 13 January 1975.

9. See Table 6, page 46, of *United States and Soviet Progress in Space: Summary Data through 1975 and a Forward Look.* Library of Congress, Congressional Research Service Paper, Washington D.C., 2 February 1976.

10. The data for 1976 have been kindly supplied to me by Dr. Charles S. Sheldon II. They differ somewhat from the 1976 figures given in *Soviet Aerospace,* Washington D.C., 8 December, page 76; and 20 December, page 124, 1976; and 10 January, page 13, 1977.

11. R. W. Kelebesadel, I. B. Strong, and R. A. Olson, Astrophys. Jr. Letters 182, (1973), L85. 188, (1974) L1.

12. *Soviet Space Programs, 1971–75,* vol. 1, chapter 1, (Washington, D.C.: U.S. Government Printing Office, 1976).

13. Dr. Charles S. Sheldon II, *U.S. Air Force Magazine,* March 1976, p. 87.

Chapter IX. Human Purpose and the Progress of Civilization

1. *Science and the Modern World* (Cambridge University Press, 1926), which includes the eight Lowell lectures of 1925. But this quotation is from Chapter 12, "Religion and Science" which, Whitehead states in the Preface, formed another address in Harvard.

2. English translation in *Moscow News,* 11 October 1975.

3. D. R. Kelley, K. R. Stunkel, and R. R. Westcott, *The Economic Superpowers and the Environment* (Berkshire, England: W. H. Freeman, 1976).

4. See, for example, *Man's Influence on Weather and Climate* by B. J. Mason, Director-General of the British Meteorological Office; Jr. Roy. Soc. Arts 125 (1977): 150.

5. Ludwig Wittgenstein, *Tractatus Logico-Philosophicus* (1921).

6. *Bulletin American Academy of Arts and Sciences,* vol. XXVIII, no. 6, page 15, 1975.

Index

About the Author

Sir Bernard Lovell, O.B.E., LL.D., D.Sc. and Fellow of the Royal Society of Great Britain, is Professor of Radio Astronomy in the University of Manchester and Director of the Experimental Station at the Nuffield Radio Astronomy Laboratories, Jodrell Bank, Macclesfield, Cheshire, England. Sir Bernard is the distinguished author of several books and many articles dealing with space exploration, the nature of the Universe, the origin of life and the nature of man's relationship to the cosmos. He also lectures throughout the civilized world on the implications of scientific growth and its consequences for man and the Universe itself.

About the Editor of This Series

Ruth Nanda Anshen, philosopher and editor, founded, plans and edits *World Perspectives, Religious Perspectives, Credo Perspectives, Perspectives in Humanism, The Science of Culture Series* and *The Tree of Life Series.* She also writes and lectures on the relationship of knowledge to the nature and meaning of man and to his understanding of and place in the universe. Dr. Anshen's book, *The Reality of the Devil: Evil in Man,* is a study in the phenomenology of evil, and is published by Harper & Row. Dr. Anshen is a member of The American Philosophical Association, The History of Science Society, and the Metaphysical Society of America.